T0205543

Springer Proceedings in Physics

Volume 211

The series Springer Proceedings in Physics, founded in 1984, is devoted to timely reports of state-of-the-art developments in physics and related sciences. Typically based on material presented at conferences, workshops and similar scientific meetings, volumes published in this series will constitute a comprehensive up-to-date source of reference on a field or subfield of relevance in contemporary physics. Proposals must include the following:

– name, place and date of the scientific meeting
– a link to the committees (local organization, international advisors etc.)
– scientific description of the meeting
– list of invited/plenary speakers
– an estimate of the planned proceedings book parameters (number of pages/ articles, requested number of bulk copies, submission deadline).

More information about this series at http://www.springer.com/series/361

Gianpaolo Carosi • Gray Rybka • Karl van Bibber
Editors

Microwave Cavities and Detectors for Axion Research

Proceedings of the 2nd International Workshop

 Springer

Editors
Gianpaolo Carosi
Lawrence Livermore National Security
Livermore, CA, USA

Gray Rybka
Physics and Astronomy Department
University of Washington
Seattle, WA, USA

Karl van Bibber
Department of Nuclear Engineering
University of California, Berkeley
Berkeley, CA, USA

ISSN 0930-8989   ISSN 1867-4941   (electronic)
Springer Proceedings in Physics
ISBN 978-3-030-06502-7   ISBN 978-3-319-92726-8   (eBook)
https://doi.org/10.1007/978-3-319-92726-8

Preface

Discovering the particles that make up dark matter is one of the major goals of physics and cosmology and one of the primary candidates is the QCD axion. Although axions are predicted to have extraordinarily weak couplings to ordinary matter, it was pointed out decades ago by Pierre Sikivie that they can be detected through their resonant conversion to photons in microwave cavities threaded by a strong magnetic field. However, only recently have experiments utilizing this technique, such as the Axion Dark Matter Experiment (ADMX), come online with the sensitivity required to probe large regions of parameter space. At their heart these experiments contain large, highly tunable microwave cavities that can be cooled to dilution refrigerator temperatures in the presence of a large magnetic field. Detecting the feeble (of order 10^{-24} W) axion-to-photon conversion signals also requires highly specialized superconducting detectors that can operate near the quantum limit (or beyond). Designing and optimizing such systems is a nontrivial task.

As a result of the increasing interest in this field, a series of workshops was organized to bring together subject matter experts in axion dark matter detection, cryogenic microwave cavity design, and quantum sensor technology in order to explore new ideas and train new researchers. These proceedings are from the "2nd Workshop on Microwave Cavities and Detectors for Axion Research," which took place at Lawrence Livermore National Laboratory (LLNL) from Jan 10 to 13, 2017. Over 40 people from around the world attended, including subject matter experts from Lawrence Livermore National Laboratory (LLNL), Stanford Linear Accelerator Center (SLAC), Lawrence Berkeley National Laboratory (LBNL), Fermi National Accelerator Laboratory (FNAL), University of Washington, University of Florida, University of California Berkeley, University of Western Australia, and Institute for Basic Science (IBS) South Korea. The proceedings are based on a series of lectures on topics ranging from the fundamentals of microwave simulations to new concepts for cavity systems and superconducting detectors. In addition, new axion detection techniques that are complementary to the standard microwave cavity search are presented.

This workshop, along with its predecessor which took place on Aug 25–27, 2015, proved useful in sharing ideas, experiences, and technologies as well as training a new generation of researchers. As a result it is anticipated that these workshops will continue on an approximately yearly basis. Both workshops and these proceedings were supported by generous contributions from the Heising-Simons foundation and was hosted at LLNL. We thank them both for their support and would also like to thank the many participants for putting this workshop and proceedings together. Finally we'd like to thank the reader who we hope will be able to use these proceedings to learn more about the exciting field of axion dark matter detection.

Livermore, CA, USA   Gianpaolo Carosi
Seattle, WA, USA   Gray Rybka
Berkeley, CA, USA   Karl van Bibber

Contents

Contributors

D. Alesini INFN, Laboratori Nazionali di Frascati, Frascati, Roma, Italy

A. Arvanitaki Perimeter Institute, Waterloo, ON, Canada

Richard F. Bradley National Radio Astronomy Observatory, NRAO Technology Center, Charlottesville, VA, USA

C. Braggio Dipartimento di Fisica e Astronomia, Padova, Italy

INFN, Sezione di Padova and Dipartimento di Fisica e Astronomia, Padova, Italy

B. M. Brubaker Yale University, Physics Department, New Haven, CT, USA

S. B. Cahn Yale University, Physics Department, New Haven, CT, USA

Gianpaolo Carosi Lawrence Livermore National Laboratory, Livermore, CA, USA

G. Carugno Dipartimento di Fisica e Astronomia, Padova, Italy

INFN, Sezione di Padova and Dipartimento di Fisica e Astronomia, Padova, Italy

Aaron Chou Fermilab, Batavia, IL, USA

N. Crescini Department of Physics and Astronomy Galileo Galilei, University of Padova, Padova, Italy

INFN, Laboratori Nazionali di Legnaro, Legnaro, Italy

N. Crisosto Department of Physics, University of Florida, Gainesville, FL, USA

M. Cunningham Department of Physics, University of Nevada, Reno, NV, USA

J. Dargert Department of Physics, University of Nevada, Reno, NV, USA

D. Di Gioacchino INFN, Laboratori Nazionali di Frascati, Frascati, Roma, Italy

Akash Dixit Department of Physics, University of Chicago, Chicago, IL, USA

H. Fosbinder-Elkins Department of Physics, University of Nevada, Reno, NV, USA

C. S. Gallo Department of Physics and Astronomy Galileo Galilei, University of Padova, Padova, Italy

INFN, Sezione di Padova, Padova, Italy

U. Gambardella INFN, Sezione di Napoli and University of Salerno, Fisciano, Italy

C. Gatti INFN, Laboratori Nazionali di Frascati, Frascati, Roma, Italy

A. A. Geraci Department of Physics and Astronomy, Northwestern University, Evanston, IL, USA

Maxim Goryachev ARC Centre of Excellence for Engineered Quantum Systems, School of Physics, University of Western Australia, Crawley, WA, Australia

M. Harkness Department of Physics, University of Nevada, Reno, NV, USA

G. Iannone INFN, Sezione di Napoli and University of Salerno, Fisciano, Italy

Yonatan Kahn Princeton University, Princeton, NJ, USA

A. Kapitulnik Department of Physics and Applied Physics, Stanford University, Stanford, CA, USA

Frank L. Krawczyk Los Alamos National Laboratory, Los Alamos, NM, USA

G. Lamanna INFN, Sezione di Pisa and University of Pisa, Pisa, Italy

S. K. Lamoreaux Yale University, Physics Department, New Haven, CT, USA

I. Lee Department of Physics, Indiana University, Bloomington, IN, USA

Y.-H. Lee KRISS, Daejeon, Republic of Korea

E. Levenson-Falk Department of Physics, Stanford University, Stanford, CA, USA

Samantha M. Lewis University of California, Berkeley, Berkeley, CA, USA

C.-Y. Liu Department of Physics, Indiana University, Bloomington, IN, USA

C. Lohmeyer Department of Physics and Astronomy, Northwestern University, Evanston, IL, USA

A. Lombardi INFN, Laboratori Nazionali di Legnaro, Legnaro, Italy

J. C. Long Department of Physics, Indiana University, Bloomington, IN, USA

Nicholas Materise Lawrence Livermore National Laboratory, Livermore, CA, USA

Ben T. McAllister ARC Centre of Excellence for Engineered Quantum Systems, School of Physics, University of Western Australia, Crawley, WA, Australia

S. Mumford Department of Physics, Stanford University, Stanford, CA, USA

A. Ortolan INFN, Laboratori Nazionali di Legnaro, Legnaro, Italy

R. Pengo INFN, Laboratori Nazionali di Legnaro, Legnaro, Italy

Nicholas M. Rapidis University of California Berkeley, Berkeley, CA, USA

G. Ruoso INFN, Laboratori Nazionali di Legnaro, Legnaro, Italy

Shriram jois Department of Physics, University of Florida, Gainesville, FL, USA

David Schuster University of Chicago, Chicago, IL, USA

Y. Semertzidis IBS Center for Axion and Precision Physics Research, KAIST, Daejeon, South Korea

Y. Shin IBS Center for Axion and Precision Physics Research, KAIST, Daejeon, South Korea

J. Shortino Department of Physics, Indiana University, Bloomington, IN, USA

P. Sikivie Department of Physics, University of Florida, Gainesville, FL, USA

E. Smith Los Alamos National Laboratory, Los Alamos, NM, USA

W. M. Snow Department of Physics, Indiana University, Bloomington, IN, USA

C. C. Speake School of Physics and Astronomy, University of Birmingham, West Midlands, UK

Ian Stern Department of Physics, University of Florida, Gainesville, FL, USA

N. S. Sullivan Department of Physics, University of Florida, Gainesville, FL, USA

D. B. Tanner Department of Physics, University of Florida, Gainesville, FL, USA

Michael E. Tobar ARC Centre of Excellence for Engineered Quantum Systems, School of Physics, University of Western Australia, Crawley, WA, Australia

E. Wiesman Department of Physics, Indiana University, Bloomington, IN, USA

Nathan Woollett Lawrence Livermore National Laboratory, Livermore, CA, USA

Sung Woo Youn Center for Axion and Precision Physics Research, Institute for Basic Science, Daejeon, South Korea

L. Zhong Yale University, Physics Department, New Haven, CT, USA

Chapter 1
Introduction to the Numerical Design of RF-Structures with Special Consideration for Axion Detector Design: A Tutorial

Frank L. Krawczyk

Abstract This publication on the numerical design of RF-structures presents two topics: (1) An overview on Numerical Methods relevant for RF-resonator design; and (2) An introduction to Simulation Software that covers 2D and 3D software tools. These include RF-design basics, introduction to Finite Difference, Finite Element and other methods, concepts for problem descriptions, interaction with particles, couplers, mechanical and thermal design, and a list of tips, tricks and challenges. This is the summary of a tutorial that the author has twice presented at a recurring "Workshop on Microwave Cavities and Detectors for Axion Research" (2nd workshop on microwave cavities and detectors for Axion Research at Lawrence Livermore Laboratory, January 10–13th, 2017, https://indico.fnal.gov/conferenceDisplay.py?confId=13068). Differences between the design of RF-structures for Axion detection and RF-structures for traditional applications are pointed out, where appropriate.

Keywords Axion · Dark matter · Simulations · Comsol · HFSS · Microwave studio · Modeling · Numerical · Cavity · Accelerator

1.1 Design Basics and Motivation for a Numerical Approach

RF-structures for practical applications, with few exceptions, cannot be designed analytically. In reflection of the wide range of applications for RF-structures, and the differences in the characteristics of EM-fields present in these structures, there are a large number of numerical design tools available that use different mathematical methods. As this tutorial shows, design tools specialize for specific applications by introducing simplifications matched to these applications, which make tools faster

F. L. Krawczyk (✉)
Los Alamos National Laboratory, Los Alamos, NM, USA
e-mail: fkrawczyk@lanl.gov

© Springer International Publishing AG, part of Springer Nature 2018   1
G. Carosi et al. (eds.), *Microwave Cavities and Detectors for Axion Research*,
Springer Proceedings in Physics 211, https://doi.org/10.1007/978-3-319-92726-8_1

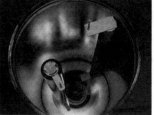

Fig. 1.1 Examples of RF structures are resonators of elliptical cross-section (left) [1] and an ADMX cavity of cylindrical cross section (right) [2]. Axion detectors are RF-resonators similar to resonators for particle acceleration. The difference is that detector cavities require tuning over a wide range of frequencies and they are not designed for a specific path of particles passing through the resonator. The Axion detector resonators also require superimposed, collinear RF and static magnetic fields in the interaction region

and simpler to use. The design tasks aim at finding geometries that can contain or transport electromagnetic (EM) fields with specific, application driven properties. Many of the more modern tools also consider secondary properties and additional conditions beyond the basic EM-properties, like mechanical or thermal properties, or the interaction with charged particles.

Resonator structures are predominantly made up of metallic enclosures of simple geometric shape that confine electromagnetic fields into volumes proportional to the wavelength of the lowest order RF fields or a fraction thereof (Fig. 1.1). The fields are thus shielded from outside environmental influences; the simple shape gives good control over field shapes, and alignment with preferred field direction. Geometric shapes are mostly derivatives of pill-boxes or enclosures of elliptical cross-section. Insertions inside these that allow guided surface current flow (quarterwave, half-wave or photonic-band-gap (PBG) structures) can be used to significantly change the size or surface losses in an enclosure. Typically multiple field solutions exist inside a resonator structure. It should be pointed out that the use of simple geometric shapes was also driven by the numerical effort to find solutions. Modern codes that benefit from faster algorithms and faster hardware also allow more complex resonator shapes, if this is beneficial to an application.

Waveguides are another class of RF-structures, used for transport of RF-fields from a source to the enclosure that are normally of rectangular cross-section, or consist of a coaxial line with metallic outside wall and a central metallic rod or pipe. For the Axion detector, coupler designs are not challenging due to the lower field levels. However attention needs to be put on minimizing perturbations to the field homogeneity inside the resonator volume.

A generic list of relevant properties that characterize RF-structures themselves includes primary properties that are a direct result of the simulation, and secondary quantities that require further processing of the primary solutions. Primary solutions include electric and magnetic field patterns (see Fig. 1.2) of a range of eigenmode solutions, and their related frequencies. The field patterns are an important criterion

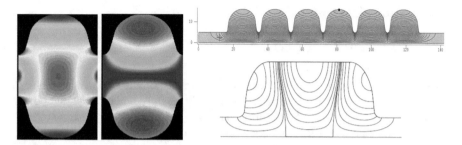

Fig. 1.2 From the left to the right: Electric and magnetic field amplitude in a single cell resonator [3]. On the top right a multi-cell, coupled resonator is shown. All these patterns are for a simple low order mode. The bottom right structure shows a higher order mode [4]. Mode order is defined by the number of sign changes a field makes in axial, radial or azimuthal direction

to select a mode of interest among the set of field solutions. Primary solutions also include peak surface field amplitudes and where they occur. For waveguides, propagation constants and field patterns can be considered primary solutions. For Axion detector cavities the operation mode will be one that can easily be aligned with a constant external magnetic field.

Secondary solutions include resonator losses and their distribution, quality factor (ratio of stored energy over these losses), coupling strength to RF-power feeds, and tuning sensitivities. Other characteristics that consider the efficiency of field interaction with an application (e.g. energy transfer from or to charged particles) include transit time factor, shunt impedance and accelerating voltage. Some of these are relevant for the Axion detector properties, but most are less relevant due to the extremely low field amplitudes and the more generic interaction with particles, as particles do not have a well-defined alignment with the field orientation, when crossing the RF-detector. An important property that is unique to the Axion detector field is the form factor. This quantity measures the alignment of the dominant RF-field with an overlapped homogeneous static magnetic field. This quantity is a dominant determining factor for the detector sensitivity. Secondary properties for waveguides include RF-losses, and the potential for a resonant field interaction with surface charges (multipacting).

RF-structures also require physical controls to introduce small changes in frequency, called tuning. This is needed to compensate fabrication variations, small frequency changes due to operation conditions, or the need for intentional sweep through a range of frequencies. Structure sensitivity to tuning is driven by mechanical properties of a resonator and the amplitude of the electric and magnetic fields close to the inside surface. Tuning sensitivity is another secondary solution type. Note that tuning for the Axion detector cavities has a more important meaning. Beyond setting and maintaining a well-defined operation point, tuning is required to sweep the resonator frequency through a wide range while searching for detector signals.

1.2   Numerical Methods

1.2.1   Reduction of Effort Strategy

Historically software has been developed to optimally simulate specific configurations in a fast, efficient and accurate manner. Only recently, when computing power has increased and numerical algorithms have become more powerful, less specialized, general purpose software tools, applicable to a wider range of applications, have been introduced. As, however, even those use the concept of reduction of effort for specific types of problems, it is worth reviewing simplification strategies.

Electromagnetic fields in free space and inside metallic enclosures are described by Maxwell's equations (Eqs. 1.1, 1.2, 1.3, and 1.4).

$$\text{Gauss's law for electric fields}: \quad \nabla \cdot E = \rho/\varepsilon_0 \tag{1.1}$$

$$\text{Gauss's law for magnetic fields}: \quad \nabla \cdot B = 0 \tag{1.2}$$

$$\text{Faraday's law}: \nabla \times E = -\frac{\partial B}{\partial t} \tag{1.3}$$

$$\text{Ampere's law}: \nabla \times B = \mu_0 \left(J + \varepsilon_0 \frac{\partial E}{\partial t} \right) \tag{1.4}$$

Solutions can be simplified or accelerated when special formulations of Maxwell's equations are used. These formulations utilize specific symmetries or simplifications of the equations based on expected field properties, upfront introduction of specific time-dependence, or symmetries of solutions. Typical examples are the solution of Poisson's equation for static electric fields (Eq. 1.5), and Helmholtz's equation for eigenvalues or time harmonic problems (Eq. 1.6), given below.

$$\nabla^2 \varphi = -\frac{\rho}{\varepsilon} \tag{1.5}$$

$$\frac{1}{c^2} \frac{\partial^2 E}{\partial t^2} - \nabla^2 \cdot E = - \left(1/\varepsilon_0 \, \nabla \rho + \mu_0 \frac{\partial J}{\partial t} \right) \tag{1.6}$$

Faraday's and Ampere's laws are only directly used for fully time dependent problems (Eqs. 1.3 and 1.4).

Numerical methods can also be formulated to consider a certain type of structure, inherent symmetries of a structure, the need to consider materials other than vacuum or metal, or the consideration of details for the RF-properties. Numerical

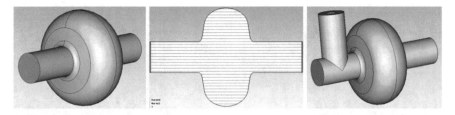

Fig. 1.3 Purely azimuthally symmetric structures (left) can be solved much faster and with higher accuracy when treated as 2D problems (center). However, simulations might require full 3D considerations, when attachments like a coupler port break the symmetry (right) [5]

methods might also be formulated to consider interaction with other secondary structures, like couplers, power feeds or tuners; or the interaction with other physics characteristics, like mechanical properties, thermal properties or the inclusion of superimposed static fields or charged particles.

Azimuthally or transitionally symmetric RF-structures can be designed much faster and with higher accuracy if these symmetries are considered in the formulation of the software already. Designers need to be aware that 2D tools can only be accurate if all aspects of the problem have the assumed symmetry. Thus symmetric structures like those shown in Fig. 1.3 can only be properly solved when no symmetry breaking features are added (Fig. 1.3 right) and when the modes of interest have the same symmetries as the structure itself. Note that symmetric structures can sustain modes that do not have the same symmetry, e.g. a deflecting dipole mode with an electric field direction normal to the central symmetry axis. Such modes cannot be calculated with a tool that assumes inherent symmetry.

1.2.2 Discretization of the Calculation Domain

Maxwell's equations describe field solution in a spatial and temporal continuum. It is difficult to write down a closed-form solution for the fields in a resonator at any point when field sources and topologies are mathematically non-trivial. Solutions for complex problems can be obtained when the global problem can be broken up into smaller problems over subvolumes in which the solution has a simple variation. Thus most numerical methods are based on a discretization of the volume of interest. A 2D or 3D grid is overlaid onto the region over which the RF-fields extend. The grid in general is inhomogeneous, selected to properly describe surface shapes and material distributions, and should lead to small enough grid elements over which the assumption of a simple variation of the solution is warranted. Figure 1.4 introduces useful features for the description of non-rectangular structures; Fig. 1.5 shows a range of modern options for discretizing volumes and structures for more accurate representation.

The simplest grid concept uses constant square grid elements and a material distribution that only knows full or empty elements. Such a simple discretization

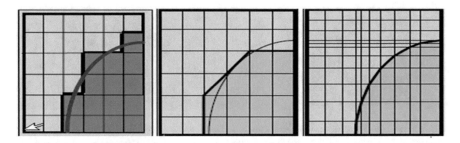

Fig. 1.4 The quality of surface and structure representation is determined by the selection of grid properties and use of simple rules for distribution of materials in grid elements [6]

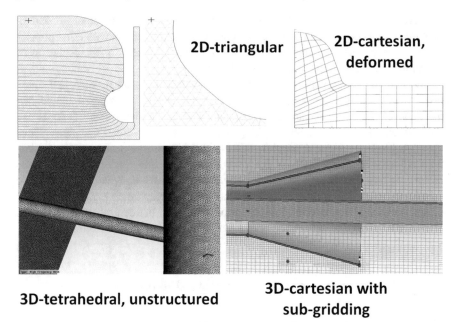

Fig. 1.5 A range of grid element choices in 2D (top row) and 3D (bottom row) that improve surface representation of modeled structures [5]

can only accurately represent rectangular structures. Curved surfaces will be approximated by a stair-step surface in such a model (Fig. 1.4, left). In 2D better approximations can be obtained, when partially filled grid elements are introduced (Fig. 1.4, center). Allowing for variable grid edges combined with partially filled cells achieves a very good approximation for curved surfaces (Fig. 1.4, right). This strategy, however, only works in two dimensions when curvature only applies in one coordinate direction. For 3D structures and general curved surfaces other strategies are required. Figure 1.5 introduces a few examples of other strategies for improved representation of general structures in discrete space.

In 2D, triangular gridding can more easily describe complex surfaces (Fig. 1.5, top, left), as can deformed rectangular grid elements (Fig. 1.5, top, right). In three dimensions, rectangular regular grid elements, even with partial fillings, have limited representation of complex geometries. Tetrahedral grid elements provide a superior geometric representation (Fig. 1.5, bottom, left). For Cartesian meshes with regular grid elements, sub-gridding can be applied for certain types of problems, which also leads to more accurate geometric descriptions.

Discretization methods used in simulation codes vary in their choice of discretization of the calculation space (as seen above), the choice of discretization of the field function, and their allocation on the discrete grid (e.g. on points, along edges or inside each discrete volume), and the consideration of surface, volume, and exclusion areas. Correct and unique solutions also need to properly consider the properties of boundaries inside the calculation region or at the outside edges of this finite calculation volume.

Beyond the proper description of complex geometries and specific problem properties, the type of equations solved and the expected size of a problem determine suitable discretization strategies. As examples, tetrahedral grid elements for general problems require more memory while providing more accurate geometry descriptions, while rectangular Cartesian elements for time domain problems are very memory efficient and support highly parallel solver algorithms. In modern codes the problem formulation often is tailored for the computer architecture to benefit from multi-core or parallel computer hardware.

1.2.3   Finite Difference Vs. Finite Element Discretization

This section describes the two most important discretization strategies and the mathematical problems they are most suitable to solve. These are the basis for the majority of simulation codes available.

The first one is the Finite Difference (FD) [7] or Finite Integration Technique (FIT) [8]. In this approach the differential or integral operators are discretized and replaced by low order difference equations. These operators are used to express Maxwell's equations in discrete form for each grid element. The result is a matrix equation that describes the full problem in discrete form. Solutions are calculated numerically applying matrix solvers (direct or iterative) suitable for the type of problem described. The discrete equations couple simple solutions from neighboring grid elements to construct field solutions across the full calculation volume of the problem. FD or FIT algorithms are typically used for rectangular Cartesian grid elements.

The strength of this approach is a simple memory efficient matrix system. Coupling to neighboring elements is similar everywhere in the calculation region (there is always a left, right, top, bottom, front and back neighbor at fixed distance). As a result only specific bands inside the matrix have non-zero elements, permitting to store a small number of linear matrix bands instead of the full matrix, which

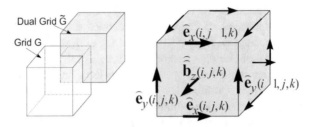

Fig. 1.6 The Yee algorithm suggests allocation of electric and magnetic fields in shifted grids. This approach allows to formulate the discrete Maxwell's equations in smooth field components only, which improves matrix condition and convergence speed for numerical solutions [9]

saves a lot of memory. The banded structure makes these techniques most suitable for solutions using vector or parallel solver strategies. Figure 1.6 shows the most common allocation of field components on the grid elements. Field components are allocated along the edges of the grid elements; thus each component in a grid element is allocated at a different location. Furthermore, electric and magnetic field components are allocated on separate grids (Fig. 1.6, left), shifted by half a grid edge length in each direction. This approach has been suggested by Yee [10], and has the advantage that the matrix systems to solve have better convergence conditions as only smooth field components are considered. The Yee allocation can also be interpreted as an allocation strategy, where the electric field components are allocated along the edges of the grid elements, while the magnetic flux components are allocated normal to the faces of each grid element (Fig. 1.6, right). If material fillings and surfaces are aligned with the grid elements, then only electric field components tangential to surfaces and magnetic field components normal to surfaces are calculated. These are the smooth components of the fields that obey Maxwell's equations.

The weakness of the FD or FIT approach is that surface representations are less accurate than in a discretization with less regular elements.

Examples of difference operators are given in Eqs. (1.7), and (1.8), (1.9). They differ in the order of coupling to neighboring grid elements. The appropriate discrete operators are determined by the solver type applied to a discrete problem.

$$\frac{\partial u}{\partial x_i} \approx \frac{u_{i+1} - u_i}{\Delta x} \quad \text{(forward difference)} \tag{1.7}$$

$$\frac{\partial u}{\partial x_i} \approx \frac{u_i - u_{i-1}}{\Delta x} \quad \text{(backward difference)} \tag{1.8}$$

$$\frac{\partial u}{\partial x_i} \approx \frac{u_{i+1} - u_{i-1}}{2\Delta x} \quad \text{(central difference)} \tag{1.9}$$

A second order derivative would have the form of Eq. (1.10).

$$\frac{\partial^2 u}{\partial x^2}_i \approx \frac{u_{i+1} - 2u_i + u_{i-1}}{(\Delta x)^2} + O(\Delta x)^2 \tag{1.10}$$

These equations demonstrate how the difference operators couple the solutions in neighboring grid elements due to common points and edges among neighboring elements. The coupling coefficients u contain the length of an edge and the materials along the edge (for electric fields) or in the normal surfaces (for magnetic fields). Contributions from different materials from all involved neighboring cells are naturally considered in this scheme. Solver algorithms for matrix systems from FD or FIT discretization tend to minimize the local energy in each grid element. Thus solutions obtained will have a known small local error; the global error of the solution is less accurately known. Another useful feature of the FIT formulation is that the difference operators in the discrete Maxwell's equations fulfill the same identities as the continuous vector-analytic operators, e.g. *curl grad F* $\equiv 0$.

The second common method of discretization is the Finite Element (FE) [11] approach. This approach has originally been developed for the solution of problems with static magnetic fields. In recent decades this approach has also been established for problems that solve for RF-fields. In the FE approach the differential or integral operators are continuous and act on discrete, low-order polynomial approximations of the field functions. Figure 1.7 shows two types of elements, on the left a tetrahedral element uses a linear polynomial representation of a field function, where the field value is represented by a linear combination of the field values at the three points that make up a face of the tetrahedron. The coupling coefficients α include the local information on the edges of the tetrahedron and on the material distribution

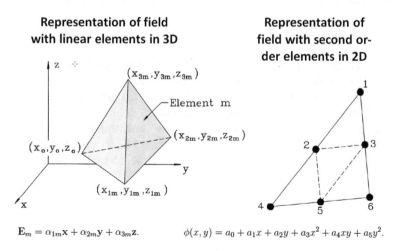

Representation of field with linear elements in 3D

$\mathbf{E}_m = \alpha_{1m}\mathbf{x} + \alpha_{2m}\mathbf{y} + \alpha_{3m}\mathbf{z}.$

Representation of field with second order elements in 2D

$\phi(x, y) = a_0 + a_1 x + a_2 y + a_3 x^2 + a_4 xy + a_5 y^2.$

Fig. 1.7 Two examples of FE type elements and optional polynomial representations of field functions [5]

along the edges. The element on the right shows a 2D triangular element, where the field for the element is represented by a second order polynomial of the field values at the corner points of the element and the values at the mid-point of each edge. Again, the coupling coefficients include the information on element size and local material distribution. Coupling between elements and construction of a field solution in the full calculation region is provided by the common points and edges between neighboring grid elements.

FE grid elements are often irregular elements, which allow a better approximation of complex structures. However, the sparseness of the matrix system and the required memory depend on the element type and on the polynomial order of the field functions. Higher polynomial order gives better accuracy, but results in more complex matrix equations. In general FE discretization gives a better representation of the problem, but the solution comes at a higher cost and is slower. Implementation of parallel solver algorithms is harder. Solver algorithms for FE type matrix problem in general minimize the global energy integral and thus give better information about the global error than about the local error from element to element.

1.2.4   Other Approaches

Depending on the application not all RF-problems can be solved efficiently with these two approaches. There are a number of alternate methods and related software tools that address specific needs more efficiently. For completeness, a list briefly summarizing some of the available methods is provided:

- *Boundary Integral Methods* or *Method of Moments*: These methods allow calculation of continuous volume solutions from sources on discretized metal surfaces.
- *Transmission Line Matrix*: This method is suitable for the solution of resonator problems as lumped circuit models.
- *Scattering Matrix Approaches*: This formulation uses a quasi-optical approximation based on wave diffraction from small features.
- There are a number of specialized solvers for the solution of fields inside conductors, e.g. metals with finite conductivity or plasmas.
- A huge challenge for numerical codes is the consideration of structures and waves of largely different length scales. To address these specialized solvers are used for examples to merge optical systems with regular RF-structures (e.g. for the calculation Smith-Purcell gratings).

1.3   Software and Design Concepts

The next part of the tutorial presented several groups of software packages that are available for RF-structure design. The list is not comprehensive but tries to cover the most common and established software packages, more software links can be found at http://www.cvel.clemson.edu/modeling/EMAG/csoft.html.

1.3.1   2D Software Tools

2D codes were the standard in simulation tools up to 10 or 15 years ago. While they are still relevant, their use is decreasing. Their strength is speed and accuracy. Their main field of application is the design of superconducting RF elliptical resonators, where peak surface fields are of the utmost importance. The most important 2D software packages and their features are listed in Table 1.1.

1.3.2   3D Software Tools

The strengths of 3D codes are the capability of treatment of complex geometries, the support of general CAD formats that allow import and export of structures for other design or evaluation tasks, and that these codes in general have more flexible, often user programmable post-processing. Many of these are commercial with professional user interfaces and design controls. However, they are slower and need much more expensive computation resources. The most important 3D software packages and their features are listed in Table 1.2.

Table 1.1 Common 2D software packages for RF-structure design

Package name	Package features
Superfish family of codes (http://laacg.lanl.gov/laacg/services/)	2D (rz, xy), FD, triangular grid elements, TM (TE) modes, RF losses, post-processing, part of general purpose suite of codes
Superlans codes [12]	2D (rz, xy), FE, quadrilateral grid elements, TM modes, RF losses, post-processing
Field precision codes (http://www.fieldp.com/)	2D (rz, xy), FE, triangular grid elements, TM/TE modes, RF losses, some post-processing, part of general purpose suite
2D modules of the MAFIA package, Urmel, TBCI (http://www.cst.com/)	2D (rz, xy), FIT, Cartesian grid elements, TM/TE modes, RF losses, post-processing, general purposes suite, particle in cell (PIC) and wake field simulations. These are not distributed anymore, but still used at many accelerator laboratories.

Table 1.2 Common 3D software packages for RF-structure design

Package name	Package features
MAFIA (http://www.cst.com/)	2D/3D (xy, rf, xyz, rfz), FIT, Cartesian grid elements, RF losses, post-processing, general purpose suite, PIC and wake field simulations. Historically, MAFIA was the first 3D general purpose package for design of accelerator structures
GdfidL (http://www.gdfidl.de/)	3D (xyz), FIT, Cartesian grid elements, RF losses, post-processing, general purpose suite, wakes, high performance computing (HPC) support
CST microwave studio (http://www.cst.com/)	3D (xyz), FIT/FE, Cartesian/tetrahedral grid elements, RF losses, post-processing, general purpose suite, PIC and wake field simulations, thermal properties, HPC support
HFSS (http://www.ansoft.com/products/hf/ hfss/)	3D (xyz), FE, tetrahedral grid elements, RF losses, post-processing, general purpose suite, interface to mechanical/thermal design tools, HPC support
Analyst (http://web.awrcorp.com/Usa/ Products/Analyst-3D-FEM-EM-Technology/)	3D (xyz), FE, tetrahedral grid elements, RF losses, post-processing, HPC support, wake field simulations
Comsol (http://www.comsol.com/)	3D (xyz), FE, tetrahedral grid elements, RF losses, post-processing, part of a multi-physics suite including mechanical/thermal and user programmable design tools
Vorpal (http://www.txcorp.com/products/ VORPAL/)	3D (xyz), FE, tetrahedral grid elements, RF losses, post-processing, particle and wake field simulations, HPC support
Remcom codes (http://www.remcom.com/)	3D (xyz), FD, Cartesian grid elements, RF losses, superior complex material models, post-processing, HPC support
SLAC ACE3P (http://www.slac.stanford.edu/ grp/acd/ace3p.html)	3D (xyz), FE, tetrahedral grid elements, RF losses, post-processing, PIC and wake field simulations, HPC support

1.3.3 RF Structure Design Concepts

1.3.3.1 Structure Description

The tables above indicate that there is a wide range of different software packages. Nevertheless, the steps of defining a structure and an overall electromagnetics problem are very similar for each package. They are based on a problem description in a discrete space, in which the RF structures are placed, and the description of the requirements to unambiguously solve a system of partial differential equations.

&po x=0.,y=0. &
&po x=0.,y=.297 &
&po nt=2,x0=0.,y0=.54071,y=-0.04232,x=.24001 &
&po x=.342,y=1.07680 &
&po x=.342,y=2.0 &
&po x=1.0253,y=3.8734 &
&po x=1.7267,y=5.7469 &
&po x=2.428,y=7.6203 &
&po x=2.428,y=7.7274 &
&po nt=2,x0=3.063,y0=7.7274,x=0.,y=.635 &
&po x=3.4638, y=8.3624 &
&po x=8.3624, y=3.4638 &
&po x=8.3624, y=3.0630 &
&po nt=2, x0=7.7274, y0=3.0630,x=0.,y=-.635 &
&po x=7.6203, y=2.428 &
&po x=5.7469, y=1.7267 &
&po x=3.8734, y=1.0253 &
&po x=2.000, y=.3420 &
&po x=1.07680,y=.342 &
&po x=.49839, y=.24001 &
&po nt=2,x0=0.54071,y0=0.,x=-.24371,y=0. &
&po x=0.,y=0. &

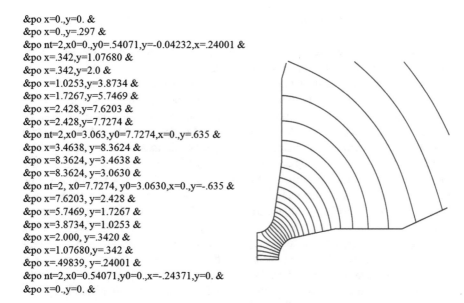

Fig. 1.8 Typical description of a 2D polygon describing a RF-structure surface: The text on the left shows a typical description of the polygon for the outline of the figure on the right. (x, y) pairs describe points; without any additional qualifier they are connected by straight lines; additional information specifies other connectors, like arcs

This tutorial section is guiding the reader through geometry description concepts, assigning material properties to the geometries, assigning boundary conditions and setting meshing and solver controls.

Solvers for 2D problems in general have simple structure definition concepts. Typically a structure is described by a closed polygon for the cross-section of the structure. Figure 1.8 provides an example of the Superfish RF design code.

Polygons are described by enumerating points on the surface and descriptors for the connections between points. These can be straight lines or curved segments. The syntax varies with the connection type. The polygon description in Fig. 1.8 is just an example; specifics are documented in syntax descriptions that come with each software tool. 2D codes, in general, require upfront calculation of specific coordinates; they do not support use of parameters or mathematical expressions. For complex geometries the use of spreadsheets for coordinate calculations is recommended.

For 3D software, some simple number based concepts do exist, but most modern codes use CAD system like structure generators. These are based on a structure assembly from structure primitives (bricks, cylinders, etc.) that are superimposed using Boolean operations. Typically the structure generators also support functions like deformation, cutting, blending, extrusions and transforming. CAD models can also often be imported or exported to or from other software.

1.3.3.2   Material Properties

There is a major difference in CAD descriptions for mechanical or thermal design and RF-design. The latter requires modeling of the inside space of a resonator enclosure, not the specific modeling of the enclosure itself. This often requires increased design efforts, as models for the enclosure and the inside have to be modeled separately. For RF simulations enclosure information is only required for RF-loss calculations and the related thermal properties. The RF-model itself will include vacuum space, dielectrics, and permeable materials. Vacuum, perfect conductors and non-lossy dielectrics are standard in all codes. Newer codes also support permeable materials and lossy properties. Very few codes, with exception of magnet design codes, can handle non-linear materials, where material properties vary with field amplitude. When losses in dielectrics and ferrites need to be considered during the simulations, they require algorithms formulated for complex numbers. Losses in metal surfaces can be considered more easily. Unless skin depths are macroscopic (very low frequencies), field penetration into the skin depth layer does not affect the solution and can be ignored. RF-codes calculate fields under the assumption of perfect conducting metals. RF-losses are later calculated in a post-processing step from the bulk resistivity and the surface magnetic fields from the field simulation.

1.3.3.3   Boundary Conditions

Unique solutions for partial differential equations, like Maxwell's equations, require knowledge about solution values or solution properties on the limiting boundaries of the finite calculation volume. This information is given by the physical problem to be solved. Common boundary conditions are Dirichlet conditions (constant potentials or vanishing tangential field values) or Neumann conditions (constant derivatives of potentials or vanishing normal fields). Very often the presence of perfect conducting enclosures provides a natural boundary condition, e.g. the electric field solution has vanishing tangential components and only non-zero normal components. Also note that electric and magnetic fields are orthogonal, so a Dirichlet condition for the electric field corresponds to a Neumann condition for the magnetic field solution. Calculation volumes are often rectangular, and software typically allows picking unique boundary conditions for each volume face.

Besides the conditions for fields in a resonator there are also other boundary conditions that RF-design software supports, where applicable. These include waveguide port and open boundary conditions. Waveguides connected to resonators can be modeled by short longitudinally invariant sections. Their terminations are modeled as an impedance-matched layer. This boundary condition is important for evaluation of resonator-coupler interaction. Open boundaries are important for simulation of solutions radiating into open space, e.g. antennas. The methods for their implementation depend on the physics problem. Methods used are expansions of solutions into their multipole moments, absorbing boundary conditions [13], or

Perfectly Matched Layers (PML), which is an improved type of absorbing condition that works better over a wide range of frequencies and range of angle of incidence of a wave when it hits a boundary.

1.3.3.4 Solver and Meshing Controls

Besides the descriptions relevant to the RF-structure, there are a number of parameters and configuration settings that need to be done before a simulation can be performed. For general purpose codes, the proper problem-type, the related solver, and both their configuration settings need to be selected. Meshing controls in most cases can be left to automeshers, but for older codes and for more complex problems proper meshing controls are relevant to describe the physics of the RF-fields correctly. Meshing controls go hand in hand with the frequencies of interest. As a rule of thumb, linear FD and FE models require a resolution of ten grid elements or more for the highest relevant frequency calculated. For higher order FE models this criterion can be reduced to four to five grid elements per wavelength. Meshing controls should also implement gradual meshing changes, where possible; the presence of materials with high dielectric of permeability properties also require increased meshing. For calculations in the time-domain convergence criteria, like the Courant stability criterion [14], need to be considered to set the largest stable time step, based on mesh size and properties of the materials in the calculation volume.

1.3.3.5 Other Relevant Features of Software Tools

The suitability of a software package also should consider flexibility, secondary processing, and speed of calculation. Flexibility is largely increased when software supports parameterization for structure description and controls. Most modern 3D tools support this. Note that the use of imported geometries from external CAD systems results in structure definitions that are static. Modern tools also support hard-wired or freely programmable optimization routines which go hand in hand with parameterization.

Most codes include basic post-processing of solutions, like graphical display and calculation of secondary properties like RF-losses and beam interaction qualifiers that are relevant for accelerator applications. This is not sufficient for a lot of applications. Modern 3D software tools often also include user programmable post-processing to derive complex solver properties as part of the solution process.

3D structures very often lead to long simulation times, thus support of modern computer hardware by the selected solver algorithms can decide if a specific tool can solve a complex problem. Modern tools are either specifically designed for support of faster hardware or include the choice for the support of such hardware. The relevant technologies to look for are multi-core CPU support, support of massively parallel computer architectures, and the support of GPU computing.

RF-designs are not stand-alone; feasibility of fabrication, mechanical stability, and thermal loads need to also be considered. Thus, it is recommended to select tools that either integrate with these evaluations or provide a simple interface to them. General purpose and multi-physics tools also have the advantage to permit evaluation of several aspects of a problem without a complete re-build for each domain of evaluation. As a reminder, note that EM fields require meshing of enclosed volume, but thermal/mechanical properties require meshing of enclosure. This needs to be considered during the structure generation.

1.4   Tip and Tricks

This final section collects a number of useful tricks for faster or higher accuracy solutions that require strategies beyond the simple modeling of a problem.

The first tip is related to the general meshing strategy. Often the basic design for a resonator is for a low order mode. Calculation of higher order modes (HOMs) is often to investigate that these do not interfere with the operation (e.g. if they are close to a harmonic of the operation mode). A fast track to a solution is to first design for the lower order mode (with decreased meshing requirements) and only run the simulations with a finer grid for the HOMs after the principle design task is completed.

Boundary conditions cannot only be used to define the properties at the outside of a problem. They can also be used to reduce the size of a problem or to enforce finding specific modes only. To achieve this, you model one half, quarter or eights of the problems and use a boundary condition to describe the field properties in symmetry planes (e.g. Fig. 1.9).

In general RF-resonators are closed structures; however there are cases, where there are significant openings in a structure. One typical application is an accelerator structure that needs to allow for the particle beam to enter and leave a structure unobstructed. Also Axion detector cavities with rotatable rods might have openings to the outside. In such cases boundary conditions are not always obvious. For power feeds, waveguide conditions might be appropriate, but there are situations that require the assumption for more open conditions. Figure 1.10 shows an elliptical resonator with a beam pipe on both ends. The geometry is designed to confine the basic RF-mode, but to let all other modes leave. For the basic design of the operation in the fundamental mode the configuration has to be selected to properly verify the confinement. This is achieved by selecting a pipe opening far enough from the structure so that the boundary conditions do not affect the confined mode. To verify this, the resonator can be calculated once with a Dirichlet and once with a Neumann boundary condition. If the frequency and all other relevant properties for this mode are not changing with the boundary condition, than these parameters are those of the confined mode.

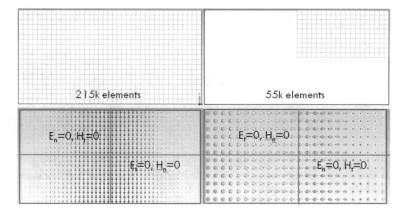

Fig. 1.9 Even for small problems like a waveguide in 2D the reduction of the problem size to one quarter is useful. Each symmetry plane can reduce the element number by a factor of two. The bottom two electric field plots show that two different sets of boundary conditions give you two different types of mode solutions [5]

Fig. 1.10 Confined modes in an "open" structure can be calculated by moving boundaries far enough away. When RF properties are independent of the selected boundary conditions the problem is properly stated [5]

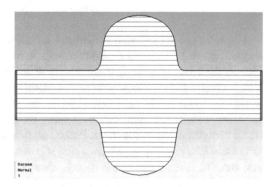

One of the hardest tasks is the modeling of changes in solutions due to small changes in the structure geometries. Defining two slightly different structures and comparing their solutions can be very inaccurate. The problem is that there is always a small discretization error due to the discrete representation of the structure. If this error is of the same order than the actual effect of the change, the simulation does not provide enough accuracy. There are a number of strategies to reduce the effect of changes in the discretization error. For tuning sensitivity, where structure changes are slight deformations of the structure surfaces, two strategies are possible. One can model a few larger deformations and check if the RF-property changes are linear. In such a case the solutions for the smaller changes can be obtained by extrapolation. A second approach is the use of a perturbation technique called Slater Perturbation Theorem ([15], see Eq. 1.11). Here no calculation of the deformed structure is done, but the expected local volume changes and the changes in stored energy due to the electric and magnetic field are used to determine a global frequency change. Increase in magnetic field energy increases the resonator frequency, while increase

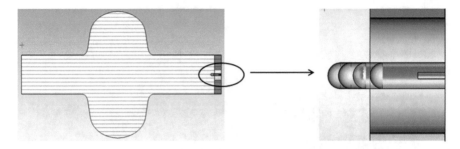

Fig. 1.11 Coupling of a field probe versus probe position can be modeled by modeling all tips at the same time and changing the length by switching assigned materials to segments between metal and vacuum [5]

in electric field energy decreases the resonator frequency.

$$\frac{\Delta\omega}{\omega} = \frac{\int \left(\mu_0 H^2 - \varepsilon_0 E^2\right) dV}{\int \left(\mu_0 H^2 + \varepsilon_0 E^2\right) dV} = \frac{\Delta U_m - \Delta U_e}{U} \tag{1.11}$$

For more predictable small changes the following strategy has also proven to be useful. Figure 1.11 shows a resonator with a weakly coupled measurement probe. Due to the weak electric field at the probe location, even for macroscopic probe length changes, the coupling would be dominated by the discretization error. To avoid this, it can be enforced that all probe lengths of interest are modeled in exactly the same grid. This is done by segmenting the probe in an overlap of all segments that are modeled at the same time. Probe length in this case is not changed by adding or removing segments, but by changing segment material properties between metal and vacuum. Thus the discretization is always the same, only the matrix system to solve is slightly changed. A similar effect can be achieved in some of the commercial 3D solvers that have learning algorithms, where within a certain topology small changes in a structure are represented by moving individual nodes in a volume grid, without changing the coupling relations to neighboring elements.

Another meshing challenge is the modeling of small gaps. It is not sufficient to properly model the two opposing gap surfaces. There must be at least one, and preferably two or three, layers of grid elements within the gap distance. Last but not least, if a solution looks suspicious or changes significantly with small changes in the structure, redo the same problem with a significantly different meshing or redo the same problem with a different simulation tool.

1.5 Summary

This document is the narrative of the tutorial that was presented both at the first and second "Workshop on Microwave Cavities and Detectors for Axion Research" held at LLNL. Its goal was to introduce novices in the field to the capabilities and proper use of RF-structure design tools. The assumptions for the detail and topic selections was that all participants had exposure the basic mathematics and physics and have been trained in researching a deeper level of information themselves.

Acknowledgments I am thanking Leanne Duffy for her thorough review of the tutorial document and for her helpful comments that added to the clarity of the descriptions.

References

1. Figure from the ILC Linear Collider Project page at http://www.linearcollider.org/ILC/What-is-the-ILC/The-project
2. ADMX Cavity with tuning rods from ADMX collaboration page at http://depts.washington.edu/admx/cavity.shtml
3. H. Edwards, D. Proch, K. Saito, Report of working group 5, in *ILC Snowmass meeting* (2005)
4. J. Billen, L. Young, Poisson/Superfish manual, LA-UR-96-1834, Rev 2006
5. F.L. Krawczyk, from LANL internal RF-class materials
6. U. van Rienen, H.-W. Glock, Methods and simulation tools for cavity design, in *Tutorial at the SRF09 Workshop*, (Dresden, Germany, 2009)
7. A. Taflove, S. Hagness, *Computational Electrodynamics: The Finite Difference Time Domain Method*, 3rd edn. (Artech House, Norwood, MA, 2005)
8. T. Weiland, M. Clemens, http://www.jpier.org/PIER/pier32/03.00080103.clemens.pdf
9. T. Weiland, On the numerical solution of Maxwell's equations and applications in accelerator physics. Part. Accel. **15**, 245–291 (1984)
10. K. Yee, Numerical solutions of initial boundary value problems involving Maxwell's equations in isotropic media. IEEE Trans. Antennas Propag. **AP-14**, 302–307 (1966)
11. FEM, Stan Humphries, http://www.fieldp.com/femethods.html
12. D.G. Myakishev, V.P. Yakovlev, Budker INP, 630090 Novosibirsk, Russia
13. G. Mur, Absorbing boundary conditions for the finite-difference approximation of the time-domain electromagnetic field equations. IEEE Trans. Electromagn. Compat. **MC-23**(4), 377–382 (1981)
14. T. Weiland, Advances in FIT, FDTD modeling, in *Review of Progress in Applied Computational Electromagnetics 18* (Monterey, CA, 2002)
15. J.C. Slater, *Microwave Electronics* (D. Van Nostrand Company, Inc., New York, 1950), p. 80

Chapter 2
Symmetry Breaking in Haloscope Microwave Cavities

Ian Stern, N. S. Sullivan, and D. B. Tanner

Abstract Axion haloscope detectors use microwave cavities permeated by a magnetic field to resonate photons that are converted from axions due to the inverse Primakoff effect. The sensitivity of a detector is proportional to the form factor of the cavity's search mode. Transverse symmetry breaking is used to tune the search modes for scanning across a range of axion masses. However, numerical analysis shows transverse and longitudinal symmetry breaking reduce the sensitivity of the search mode. Simulations also show longitudinal symmetry breaking leads to other undesired consequences like mode mixing and mode crowding. The results complicate axion dark matter searches and further reduce the search capabilities of detectors. The findings of a numerical analysis of symmetry breaking in haloscope microwave cavities are presented.

Keywords Axion · Dark matter · Simulations · Comsol · Numerical · Cavity · Mode structure · Mode crossings · Symmetry breaking

2.1 Background

The axion particle, first theorized as a solution to the charge conjugation and parity symmetry problem of quantum chromodynamics [1–3], has been established as a prominent cold dark matter (CDM) candidate [4]. The most sensitive axion search technique is the haloscope detector [5, 6], proposed by Sikivie [7]. The haloscope detector uses a microwave cavity permeated by a strong magnetic field to convert axions to photons via the inverse Primakoff effect [8]. The scan mode of the cavity is tuned across a frequency range to search for CDM axions.

I. Stern (✉) · N. S. Sullivan · D. B. Tanner
University of Florida, Gainesville, FL, USA
e-mail: ianstern@ufl.edu

© Springer International Publishing AG, part of Springer Nature 2018
G. Carosi et al. (eds.), *Microwave Cavities and Detectors for Axion Research*,
Springer Proceedings in Physics 211, https://doi.org/10.1007/978-3-319-92726-8_2

The power measured in the cavity for a specific resonant mode is given by [9]

$$P_{mnp} \approx g_{a\gamma\gamma}{}^2 \frac{\rho_a}{m_a} B_0{}^2 V \; C_{mnp} \; Q_L, \tag{2.1}$$

where the indices m, n, and p identify the mode. The axion parameters are given by $g_{a\gamma\gamma}$, the axion-photon coupling constant, m_a, the mass of the axion, and ρ_a, the local mass density. The parameters of the detector are given by B_0, the magnetic field strength, V, the volume of the cavity, and Q_L, the loaded quality factor of the cavity (assumed to be less than the kinetic energy spread of the axion at the Earth). C_{mnp} is the normalized form factor describing the coupling of the axion conversion to a specific cavity mode, derived from the Lagrangian for the axion-photon interaction. It is given by

$$C_{mnp} \equiv \frac{\left| \int d^3 x \; \mathbf{B_0} \cdot \mathbf{E_{mnp}} \left(\mathbf{x} \right) \right|^2}{B_0{}^2 V \int d^3 x \; \varepsilon \left(\mathbf{x} \right) \left| \mathbf{E_{mnp}} \left(\mathbf{x} \right) \right|^2}, \tag{2.2}$$

where $\mathbf{E_{mnp}}$ is the electric field of the mode an ε is the permittivity within the cavity normalized to vacuum.

2.2 Microwave Cavity Theory

When the permeability in a cavity is homogenous, the modes must satisfy [10]

$$\left(\nabla^2 + 2\pi \; \mu\varepsilon \; f^{\;2} \right) \begin{Bmatrix} \mathbf{E} \\ \mathbf{B} \end{Bmatrix} = \begin{Bmatrix} \nabla \left(\nabla \cdot \mathbf{E} \right) \\ \left(\nabla \times \mathbf{B} \right) \times \frac{\nabla \varepsilon}{\varepsilon} \end{Bmatrix}, \tag{2.3}$$

where f is the mode frequency, and μ and ε are the permeability and permittivity within the cavity, respectively. If the permittivity inside the cavity is also constant, Eq. (2.3) reduces to the eigenvalue problem [11].

$$\left(\nabla_t^2 + 2\pi \; \mu\varepsilon f^{\;2} \right) \begin{Bmatrix} \mathbf{E} \\ \mathbf{B} \end{Bmatrix} = 0. \tag{2.4}$$

For cylindrical cavities, the modes are standing waves consisting of an oscillating field with a constant cross-section, traversing along the longitudinal (z) axis. The result is three sets of infinite orthogonal modes, transvers electric (TE), transverse magnetic (TM), and transverse electromagnetic (TEM). The orthogonality applies to the field types independently ($\mathbf{E_i} \cdot \mathbf{E_j}$ and $\mathbf{B_i} \cdot \mathbf{B_j}$) as well as combined ($\mathbf{E_i} \cdot \mathbf{B_j}$).

If the permittivity inside the cavity is not constant along the z-axis, the modes are standing waves with varying cross-section, breaking longitudinal symmetry. Specifically, when the permittivity has discontinuities, such as internal boundaries, the orthogonality of the field types independently is broken [12]. The orthogonality breaking gives modes that are not pure TM, TE, or TEM, but some mixed or localized mode. These modes can have significant effects on the sensitivity and search capabilities of a haloscope detector.

Equation (2.2) shows that the sensitivity of a haloscope detector is strongly dependant on the relationship between the static magnetic field and the orientation of the electric field of the search mode. Typically, a haloscope detector uses a circular-cylinder shaped microwave cavity and a solenoidal superconducting magnet. In this configuration, the search mode would need sufficient electric field pointed in the z-axis to observe the axion-to-photon conversion.

In a circular-cylinder microwave cavity with constant permittivity, only the TM_{0n0} modes would couple to the axion, with the TM_{010} mode having the strongest coupling (C). However, because the cavity in an axion haloscope must be tuned to different frequencies to conduct a search, discontinuities in permittivity are almost inevitable. To date, haloscopes are tuned using one or more conducting or dielectric rods that run the length of the cavity in the z-direction [13]. The rods are moved transversely to change the frequency of the TM_{0n0} modes. Because the rods are moved, a physical gap must exist between the rod-ends and the endcaps of the cavity. The gaps create a discontinuity in the permittivity in the longitudinal direction, breaking mode orthogonality of the TM_{0n0} modes. This phenomenon has been identified as a capacitance effect [14].

Therefore, search modes in a haloscope detector are perturbations of a TM_{0n0} mode and frequently a mixed mode. The mixing of TM with TE and TEM modes causes gaps in the frequency scan range of a haloscope. Further, additional modes are formed due to the symmetry breaking, increasing the number of modes at similar frequency to the search mode. This mode crowding increases the difficultly of tracking a search mode, making axion detection more challenging.

2.3 Numerical Analysis

In an effort to quantify the effects of symmetry breaking on search capabilities of a haloscope detector, numerical analysis was conducted to evaluate the impact on form factor (C), frequency scan range, and mode crowding. The simulations were conducted with a commercially available three-dimensional finite element modeling program (COMSOL version 5.1). All cavity models had a diameter of 5.375 in. and a height of 10.75 in., and used a single tuning rod of diameter 1.430 in. The maximum mesh size was no more than 1/7th the wavelength and an eigenfrequency solver was used to compute the modes.

Simulations of transverse symmetry breaking showed that the form factor is decreased when symmetry is broken. The result matches similar findings in other numerical analyses [13, 15]. More significantly, the simulations showed that transverse symmetry breaking did not break mode orthogonality, and did not result in mode mixing or an increase in mode crowding about the search mode frequency.

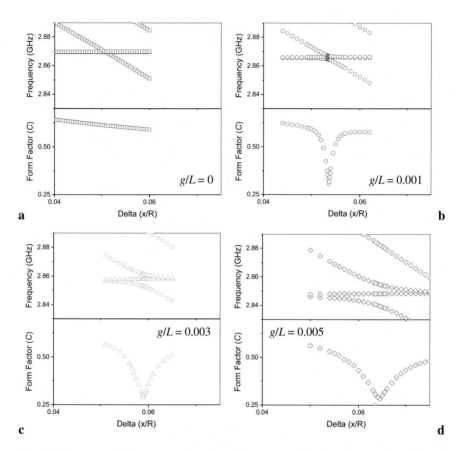

Fig. 2.1 Frequency of the lowest search mode and the corresponding form factor as the tuning rod is moved through a mode crossing of the cavity for a rod-end gap of (**a**) $g/L = 0$; (**b**) $g/L = 0.001$; (**c**) $g/L = 0.003$; (**d**) $g/L = 0.005$. The distance from the center of the tuning rod to the center of the cavity, x/R, is shown on the x-axis. As the gap increases, greater mode mixing is observed and the dip in form factor broadens. The next higher TM (non-searchable) mode is also shown

Simulations of longitudinal symmetry breaking yielded more valuable information. Adding a gap between the tuning rod and endcaps showed mode orthogonality breaking, which caused significant mode mixing at the point where the search mode had a similar frequency to another mode (i.e., a mode crossing). The mixing caused a gap in the frequency scan range of the search mode. The mixing increased as the gap grew, resulting in a larger gap in the frequency scan range.

Figure 2.1a–d shows the frequency of the lowest search mode and the corresponding form factor as the tuning rod is moved from the center of the cavity for gap-to-height ratios (g/L) of 0, 0.001, 0.003, and 0.005, respectively. The distance from the center of the tuning rod to the center of the cavity, x, is shown on the x-axis, and normalized by the radius of the cavity, R.

The plots demonstrate the effects of a mode crossing and shows a degenerate TE mode and the next higher TM mode ($C \approx 0$). The symmetry breaking causes the TE modes to break degeneracy. When the search mode is tuned to a frequency nearly the same as the TE modes, the search mode mixes with one of the TE modes, forming two mixed modes. The other TE mode does not mix with the search mode and maintains orthogonality as well as a constant frequency.

Mode mixing is depicted as a reduction in form factor and a frequency separation in modes at the crossing. The non-mixing TE mode is depicted as the straight-line data points between the frequency separation in Fig. 2.1b–d. As the rod-end gap increases, the frequency spread of the mode mixing increases, as depicted by the broadening of the dip in form factor, and the increased separation in frequency between the modes at the crossing [16]. When the gap is zero, there is no mode mixing as seen in Fig. 2.1a by an absence of a dip in the form factor at the mode crossing.

Counterintuitively, the quality factor of the modes does not decrease with the increase in gap size, indicating the mode bandwidths do not increase to account for the frequency spread. Instead, the larger rod-end gap increases the electric potential due to the capacitance effect, causing the mixing to start to occur at a great frequency difference between the two modes.

Figure 2.2 shows a cross-section of the electric field in the mixed modes during the mode crossing. The field strength is indicated by the blue area with a stronger field corresponding to lighter color. The white area indicates the tuning rod or cavity boundaries where there is no field. The red arrows depict the field vectors. Orthogonality breaking is observed. Both modes have some electric field in the z-axis, and thus neither is a TE mode. Maxwell's equations can be used to show neither mode is TM, thus they are both mixed modes. The form factors of the modes are approximately the same.

The frequency separation at the mode crossing causes a gap in the frequency scan range of the cavity. At the rod orientation where the form factor is lowest in a mode crossing, the form factors of the mixed modes are the same. At that point, a search must move from scanning one mode to scanning the other, resulting in a gap in frequency. As the separation increases, the frequency gap increases. Figure 2.3 shows the gap in frequency scan range, Δf, normalized by the mode frequency, as a function of rod-end gap. At small gap sizes, $\Delta f / f \approx g/L$.

Longitudinal symmetry breaking induces additional modes from degeneracy breaking and mode localization. The simulation showed reentrant modes appear in the rod-to-endcap gaps. The mode crowding interferes with tracking a search mode. Since the form factor of a mode is not directly measurable, the signal from the search mode is indistinguishable from signals from other modes with nearly the same frequency, complicating searches with haloscope detectors. Additionally, mode crowding increases mode crossings and mode mixing, further degrading the capability of a detector.

Figure 2.4 shows the nearest ten modes to the lowest search mode with the tuning rod along the center axis for various rod-end gaps, depicting the mode crowding as the modes move closer together with increased gap size. The modes cluster in

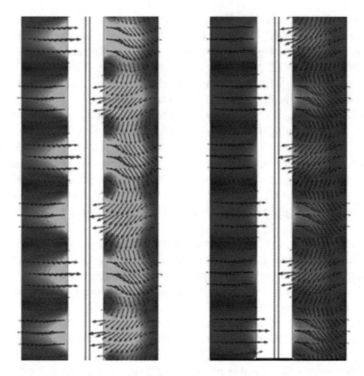

Fig. 2.2 Cross-section of electric field of the mixed modes during mode crossing. Two modes are shown with the center white area of each being the location of the tuning rod (no field). The blue area is the field strength, with lighter color indicating stronger field. The cavity wall and endcaps would touch the outer edge of the blue areas. The red arrows show the electric field vectors. The left side of both modes demonstrates a TE-like field, while the right side of both modes contain some z component of the electric field. The form factor of both modes is approximately the same

Fig. 2.3 Gap in frequency scan range due to mechanical gap between the rod-ends and the cavity endcaps. The longitudinal symmetry breaking breaks the mode orthogonality, causing modes to mix when they are nearly the same frequency. An axion search must transition from one mode to another at a mode crossing, resulting in a gap in the frequency scan range. For a single tuning rod of radius $r/R \approx 0.25$, $\Delta f/f \approx g/L$ at small gaps

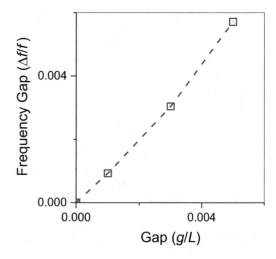

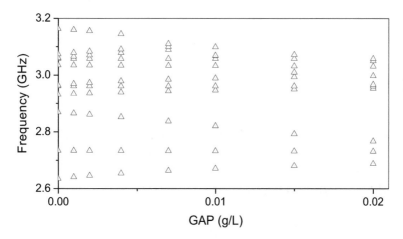

Fig. 2.4 Mode crowding induced by mechanical gaps between rod-ends and endcaps. The tuning rod is located along the center axis. As the gap is increased, modes move closer together and more modes are observed due to localization and degeneracy breaking. The plot shows two distinct clustering of modes; the search mode is found in the higher-frequency cluster

two distinct groups. Several new modes are observed as the gap increases (i.e., the highest frequency mode shown at $g/L = 0.007$ is a reentrant mode not observed at smaller gap sizes; two modes shown in the upper cluster at $g/L = 0.02$ are degeneracies not observed at smaller gap sizes).

Tilting of the tuning rod is a more complex symmetry breaking, as it breaks both transverse and longitudinal symmetry simultaneously. Mode localization can take on several forms and has been observed to produce degeneracy breaking in modes that were not degenerate prior to the symmetry breaking. Figure 2.5a–d shows the frequency of the lowest search mode and the corresponding form factor as the tuning rod is moved from the center of the cavity for tuning rod tilt (φ) of 0.25°, 0.50°, 1.00°, and 1.80°, respectively. The distance from the center of the tuning rod to the center of the cavity, x, is shown on the x-axis, and normalized by the radius of the cavity, R. Note, Fig. 2.1a shows the results for zero tilt.

The plots demonstrate the effects of a mode crossing and shows a degeneracy TE mode and the next higher TM-like mode. Mode mixing is depicted as a reduction in form factor and a frequency separation in modes at the crossing. Similar phenomena are observed as with the rod-end gaps. The mode mixing occurs within a smaller frequency span than from rod-end gaps, but the reduction in form factor across the scan range is more significant. Mode crowding was also observed, though less severe than with rod-end gaps.

At a tilt above $\sim1.00°$, higher-order tuning modes obtain a greater form factor than the lowest tuning mode during part of the scan range. At a tilt of 1.80°, the next higher TM-like mode increased frequency initially and then obtains a higher form factor ($C \approx 0.3$) approximately when the lowest tuned mode begins to mix at the mode crossing. This effect produces several large gaps in the scan range.

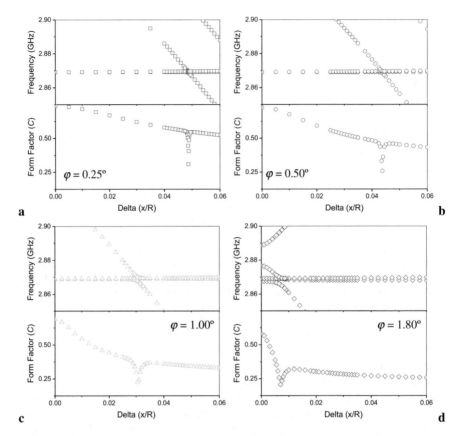

Fig. 2.5 Frequency of the lowest search mode and the corresponding form factor as the tuning rod is moved through a mode crossing of the cavity for a tilt of (**a**) $\varphi = 0.25°$; (**b**) $\varphi = 0.50°$; (**c**) $\varphi = 1.00°$; (**d**) $\varphi = 1.80°$. The distance from the center of the tuning rod to the center of the cavity, x, is shown on the x-axis, and normalized by the radius of the cavity, R. As the tilt increases, more mode mixing is observed and the form factor is further lowered across the scan range. The next higher TM-like (non-searchable) mode is also shown for completeness

2.4 Conclusion

The numerical analysis presented here showed orthogonality breaking in microwave cavity modes is a result of longitudinal symmetry breaking and is the cause of mode mixing. As the symmetry breaking increases, mode mixing, mode crowding, and the gap in frequency scan range increase, reducing the effectiveness of a haloscope detector. For the geometry analyzed, the ratio of the frequency gap to the mode frequency ($\Delta f/f$) was approximately equal to the ratio of the rod-end gap to the cavity height (g/L) for small gaps. Rod tilting caused complex mode mixing. At a tilt above $\sim 1.00°$, the most efficient search mode changed depending on rod orientation and multiple gaps in frequency scan range are observed.

Acknowledgements This research was supported by DOE grant DE-SC0010296.

References

1. R. Peccei, H. Quinn, Phys. Rev. Lett. **38**, 1440 (1977)
2. S. Weinberg, Phys. Rev. Lett. **40**, 223 (1978)
3. F. Wilczek, Phys. Rev. Lett. **40**, 279 (1978)
4. J. Ipser, P. Sikivie, Phys. Rev. Lett. **50**, 925 (1983)
5. I. Stern, AIP Conf. Proc. **1604**, 456 (2014)
6. B.M. Brubaker, L. Zhong, Y.V. Gurevich, S.B. Cahn, S.K. Lamoreaux, M. Simanovskaia, J.R. Root, S.M. Lewis, S. Al Kenany, K.M. Backes, I. Urdinaran, N.M. Rapidis, T.M. Shokair, K.A. van Bibber, D.A. Palken, M. Malnou, W.F. Kindel, M.A. Anil, K.W. Lehnert, G. Carosi, Phys. Rev. Lett. **118**, 061302 (2017)
7. P. Sikivie, Phys. Rev. Lett. **51**, 1415 (1983)
8. H. Primakoff, Phys. Rev. **81**, 899 (1951)
9. P. Sikivie, Phys. Rev. D **32**, 2988 (1985)
10. H. Zucker, G.I. Cohn, IRE Trans. Microw. Theory Tech. **10**, 202 (1962)
11. J. Jackson, *Classical Electrodynamics*, 3rd edn. (Wiley, Hoboken, 1999), pp. 363–374
12. W.C. Chew, *Lectures on Theory of Microwave and Optical Waveguides*, (unpublished), p. 96, http://wcchew.ece.illinois.edu/chew/course/tgwAll20121211.pdf
13. I. Stern, A.A. Chisholm, J. Hoskins, P. Sikivie, N.S. Sullivan, D.B. Tanner, G. Carosi, K. van Bibber, Rev. Sci. Instrum. **86**, 123305 (2015)
14. D. Lyapustin, *An improved low-temperature RF-cavity search for dark-matter axions*, Ph.D. Thesis, Univ. of Wash., Seattle WA, 2015
15. C. Hagmann, P. Sikivie, N.S. Sullivan, D.B. Tanner, S.I. Cho, Rev. Sci. Instrum. **61**, 1076 (1990)
16. C. Hagmann, *A search for cosmic axions*, Ph.D. Thesis, Univ. of Flor., Gainesville FL, 1990

Chapter 3
Pound Cavity Tuning

Shriram Jois, N. S. Sullivan, and D. B. Tanner

Abstract A Pound reflection locking method is used to tune to a desired frequency high-Q microwave resonant cavities to be used for axion detection. A phase-modulated RF signal is fed to the cavity and the reflected RF signal is detected with a zero-bias Schottky diode detector. The detected signal is analyzed using a lock-in amplifier synchronized to the phase modulation frequency. The lock-in output is integrated and amplified to provide the feedback signal to a servomechanism to lock the resonant cavity to the desired frequency.

Keywords Axion · Pound locking · Multicavity · Frequency locking · Reflection · Phase modulation · Feedback · Servomechanism

3.1 Introduction

The Peccei-Quinn solution to the strong CP problem in QCD [1] predicts the existence of Goldstone bosons called axions which, if they exist, are a possible candidate for cold dark matter [2]. These axions decay into photons in a static magnetic field due to the Primakoff effect and this is the method used by the Axion Dark Matter eXperiment (ADMX) [3] to search for galactic-halo axions using RF cavities. Because the axion mass is unknown, the search requires the cavity to be slowly tuned through the range of masses allowed by astrophysical constraints. By adjusting the tuning mechanism, the resonant frequency of the RF cavity is tuned to the frequency that matches the axion mass m_a ($f = m_a c^2/h$). At present ADMX uses a single cylindrical RF cavity which is satisfactory for low frequencies (<1 GHz). In order to maximize detector volume available in the magnet and extend the search range to frequencies above 1 GHz, a multiple cavity

S. Jois (✉) · N. S. Sullivan · D. B. Tanner
Department of Physics, University of Florida, Gainesville, FL, USA
e-mail: ramjois@ufl.edu; sullivan@phys.ufl.edu; tanner@phys.ufl.edu

© Springer International Publishing AG, part of Springer Nature 2018   31
G. Carosi et al. (eds.), *Microwave Cavities and Detectors for Axion Research*,
Springer Proceedings in Physics 211, https://doi.org/10.1007/978-3-319-92726-8_3

array has been proposed. A mechanism is required to lock these cavities together to identical resonant frequencies.

3.2   Block Diagram

Figure 3.1 shows the block diagram of the Pound circuit [4]. The RF oscillator A generates the carrier frequency (Ω).

$$V = V_0 e^{-i(\Omega t + \phi)}.\tag{3.1}$$

The oscillator is phase modulated with a modulation frequency ω generated by the lock-in amplifier, making

$$\Phi = \beta \cos(\omega t + \Phi_0),\tag{3.2}$$

and

$$V = V_0 e^{-i(\Omega t + \beta \cos \omega t + \Phi_0)} = V_0 e^{-i(\Omega t)} e^{-i \beta \cos \omega t} e^{-i \Phi_0}.\tag{3.3}$$

For a modulation index, $\beta \ll 1$, we can approximate the middle exponential term:

$$e^{-i \beta \cos \omega t} \approx 1 - i \beta \cos \omega t \approx 1 - i \frac{\beta}{2}(e^{i \omega t} + e^{-i \omega t}),\tag{3.4}$$

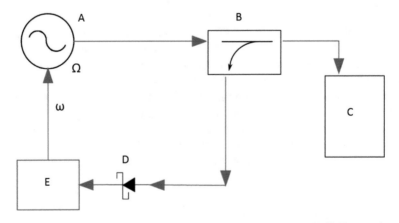

Fig. 3.1 Block diagram of the test circuit for of Pound modulation. In this figure: A = Source, providing a phase modulated carrier frequency Ω; B = Directional coupler; C = Cylindrical TM$_{010}$ cavity; D = Schottky diode; E = Lock-in amplifier. The internal oscillator of the lock-in emits a reference frequency ω which is used to phase modulate Ω

so that

$$V = V_0 e^{-i\Phi_0}[e^{-i\Omega t} - i\frac{\beta}{2}(e^{-i(\Omega+\omega)t} + e^{-i(\Omega-\omega)t})].\tag{3.5}$$

Equation (3.5) represents a phase modulated (PM) wave with a carrier frequency Ω and sidebands at $\Omega + \omega$ and $\Omega - \omega$. This PM signal is fed to the cavity using a directional coupler. The output port of the cavity is terminated with a $50\,\Omega$ terminator. The reflected signal is picked off by the weak port of the directional coupler and detected using a Schottky diode that operates as a square law detector. The detected signal is fed to a lock-in detector that is synchronized to the modulation frequency. The output of the lock-in amplifier provides dc signals that can be used to drive the cavity and tune it automatically.

3.3 Transmission and Reflection Spectra

Figure 3.2 shows the transmission and reflection spectra of the cavity used. The resonant frequency of the cavity is 1.483 GHz and the cavity Q is 13,000.

3.4 Error Signals

Figure 3.3 shows the equivalent circuit of the cavity in terms of a series RLC circuit with mutual inductance M representing the coupling. Z_i is the input impedance of this circuit and is given by Eq. (3.6) [5]

$$Z_i = i\omega L_i + \frac{\omega^2 M^2}{R_c + i\omega(L_c - \frac{1}{C_c})}\tag{3.6}$$

The reflection coefficient Γ is defined as,

$$\Gamma = \frac{Z_L - Z_i}{Z_L + Z_i},\tag{3.7}$$

where $Z_L = 50\,\Omega$ is the impedance of the RF cable. Equation (3.6) implies that Γ is a function of frequency. Substituting Eq. (3.6) in (3.7) we get,

$$\Gamma(\Omega) = \frac{Z_L - \left[i\Omega L_i + \frac{\Omega^2 M^2}{R_c + i\Omega\left(L_c - \frac{1}{\Omega^2 C_c}\right)}\right]}{Z_L + \left[i\Omega L_i + \frac{\Omega^2 M^2}{R_c + i\Omega\left(L_c - \frac{1}{\Omega^2 C_c}\right)}\right]} = \frac{Z_L - \left[i\Omega L_i + \frac{\Omega^2 M^2}{R_c\left\{1 + iQ\left(1 - \frac{\omega_0^2}{\Omega^2}\right)\right\}}\right]}{Z_L + \left[i\Omega L_i + \frac{\Omega^2 M^2}{R_c\left\{1 + iQ\left(1 - \frac{\omega_0^2}{\Omega^2}\right)\right\}}\right]}.\tag{3.8}$$

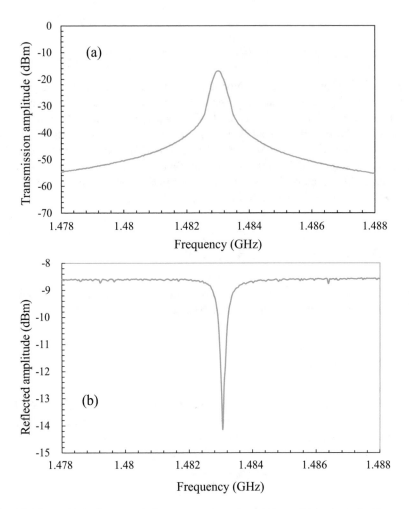

Fig. 3.2 Figure (**a**) is the transmission power spectrum and figure (**b**) is the reflection power spectrum of the cavity. The resonant frequency of the cavity is 1.483 GHz

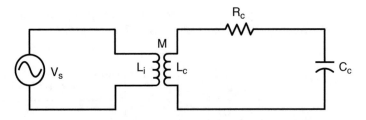

Fig. 3.3 Equivalent lumped-circuit representation of the cavity. The mutual inductance M provides the coupling to a series RLC resonant circuit

Using $\tan\theta = Q(1 - \frac{\omega_0^2}{\Omega^2})$, Eq. (3.8) can be written as,

$$\Gamma(\Omega) = \frac{Z_L - (i\Omega L_i + \frac{\Omega^2 M^2}{R_c}\cos\theta e^{-i\theta})}{Z_L + (i\Omega L_i + \frac{\Omega^2 M^2}{R_c}\cos\theta e^{-i\theta})}. \tag{3.9}$$

Because of this frequency dependence, V_{refl} will contain $\Gamma(\Omega)$, $\Gamma(\Omega+\omega)$, $\Gamma(\Omega-\omega)$ corresponding to carrier, upper side band and lower side band.

$$V_{refl} = V_0 e^{-i\Phi_0}[\Gamma(\Omega)e^{-i\Omega t} - i\frac{\beta}{2}(\Gamma(\Omega+\omega)e^{-i(\Omega+\omega)t} + \Gamma(\Omega-\omega)e^{-i(\Omega-\omega)t})] \tag{3.10}$$

Component D in Fig. 3.1 is a square law detector and produces an output proportional to $|V_{refl}|^2$.

$$|V_{refl}|^2 = V_0^2\beta[\cos(\omega t)\Re\chi + \sin(\omega t)\Im\chi + O(2\omega t) + \text{constant}] \tag{3.11}$$

where

$$\chi = \Gamma(\Omega)\Gamma^*(\Omega+\omega) - \Gamma^*(\Omega)\Gamma(\Omega-\omega) \tag{3.12}$$

We find that the components at frequency ω are given by

$$|V_{refl}|^2 = V_0^2\beta[\cos(\omega t)\Re\chi + \sin(\omega t)\Im\chi] \tag{3.13}$$

Depending on its phase setting, the lock in amplifier can select either the $\cos(\omega t)$ or $\sin(\omega t)$ term[6] referred to as the I and Q components, respectively. A two-phase lock-in can detect both. Figure 3.4 shows the two phases of the lock-in output as the frequency Ω is swept over 10 MHz passing through the cavity resonance. There is a sweep range of $\sim$100 kHz (approximately Ω_0/Q) where the error signal is a linear function of the frequency offset from resonance.

3.5 Locking Multiple Cavities

The next step is to use either I or Q as the error signal and design a drive to lock the cavity to the source. After proving that the cavity signal can be locked to the source frequency, the next step is to lock multiple cavities to the source using the same method. The drive would consist of a PI controller or a μ-controller with a motor that would zero the error signal and bring the cavity to resonance (Fig. 3.5).

Fig. 3.4 Lock-in error signal versus the generator frequency Ω near the resonance. The reflected voltage amplitude is in μV and the frequency is in MHz. I is the in phase component and Q is the quadrature component. Both I and Q have peaks at Ω, $\Omega + \omega$ and $\Omega - \omega$. The peaks at Ω pass through zero when Ω passes through the cavity resonance frequency

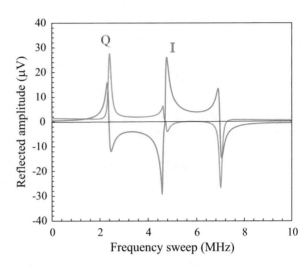

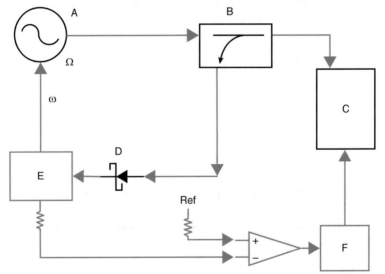

Fig. 3.5 Block diagram of a Pound circuit feedback-driven servomechanism. In this figure: A = Source, providing a phase modulated carrier frequency Ω; B = Directional coupler; C = Cylindrical TM_{010} cavity; D = Schottky diode; E = Lock-in amplifier; F = Servomechanism

References

1. R.D. Peccei, H.R. Quinn, CP conservation in the presence of pseudoparticles. Phys. Rev. Lett. **38**(25), 1440 (1977)
2. J.R. Primack, D. Seckel, B. Sadoulet, Detection of cosmic dark matter. Annu. Rev. Nucl. Part. Sci. **38**(1), 751–807 (1988)
3. P. Sikivie, Experimental tests of the invisible axion. Phys. Rev. Lett. **51**(16), 1415 (1983)

4. R.V. Pound, Electronic frequency stabilization of microwave oscillators. Rev. Sci. Instrum. **17**(11), 490–505 (1946)
5. C.G. Montgomery, R.H. Dicke, E.M. Purcell, *Principles of Microwave Circuits*, vol. 25 (IET, London, 1948)
6. E.D. Black, An introduction to Pound-Drever-Hall laser frequency stabilization. Am J. Phys. **69**(1), 79–87 (2001)

Chapter 4
Modification of a Commercial Phase Shifter for Cryogenic Applications

Richard F. Bradley

Abstract Conventional radio frequency matching techniques using silicon-based phase shifters operate well at ambient temperatures, but will perform poorly in cryogenic environments due to carrier freeze-out. However, varactors made from gallium arsenide retain carrier mobility at such temperatures. This paper describes the modification and evaluation of a commercial phase shifter operating in the UHF frequency range for cryogenic applications.

Keywords Cryogenic · Coupling · Phase shifter · Electronic · Silicon-based · Germanium-based · Axion · Commercial · Impedance matching · Antenna · Dark matter

4.1 Introduction

This work was motivated by the need for improving the radio frequency (rf) impedance match at the output port of a critically coupled cavity resonator to a transmission line, a generic representation of which is illustrated schematically in Fig. 4.1 [1]. At present, cryogenic high-Q cavities used in the Axion Dark Matter Experiment (ADMX) [2] reply on physically adjusting antenna positions to impedance match. Mechanical rods positioned on cams located within the cavity adjust its fundamental resonant frequency, but the port impedance may contain a frequency dependent reactance. A single-stub arrangement at the port can help improve power coupling to the transmission line by resonating the cavity's reactance near the port. Conventional applications make use of an rf short that slides along the stub transmission line to create this reactance. However, since mechanical motion is required, this approach is difficult to implement reliably at cryogenic temperatures. An electronic means of adjusting this reactance is desired.

R. F. Bradley (✉)
National Radio Astronomy Observatory, NRAO Technology Center, Charlottesville, VA, USA
e-mail: rbradley@nrao.edu

© Springer International Publishing AG, part of Springer Nature 2018   39
G. Carosi et al. (eds.), *Microwave Cavities and Detectors for Axion Research*,
Springer Proceedings in Physics 211, https://doi.org/10.1007/978-3-319-92726-8_4

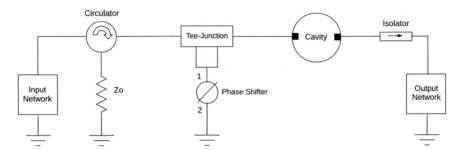

Fig. 4.1 Diagram of the basic tuning arrangement for matching a critically-coupled cavity resonator

An equivalent circuit of an electronic phase shifter that makes use of varactors is shown in Fig. 4.2 There are basically two components to the phase shifter circuit. The quadrature hybrid junction is a four-port device that divides the signal entering its port A into equal magnitudes at ports B and C with a 90° phase shift between them while providing no signal at port D (180° out-of-phase). A signal entering the hybrid's port D is split between ports B and C while providing no signal at port A. In fact, the other ports behave in the similar manner, yielding a four-port symmetry in the transfer functions. The second component of the phase shifter is the matched pair of varactors that terminate hybrid ports B and C with respect to ground. The voltage that controls the capacitance of the varactors (reversed biased diodes) is introduced through resistors R1 and R2. Capacitor C1 bypasses the external bias connection preventing rf energy from coupling onto the bias cabling.

A signal entering port 1 of the phase shifter is directed toward the hybrid junction where it is split with equal magnitudes but in phase quadrature at ports B and C. The reactance of the varactors cause these signals to reflect back into the hybrid with a given phase shift while maintaining the 90° differential phase between them. These signals encounter another phase shift by the hybrid junction resulting in both signals appearing out-of-phase at port A but in-phase at port D. If port 2 is terminated with a short, the signal is reflected back into the phase shifter and the process is reversed, producing the reflected signal back to port 1. Thus, at port 1, the maximum range of the phase of the reflected signal is between 0° and 180°, limited by the characteristics of the varactors, the bias voltage range, and the signal frequency.

Commercial phase shifters that make use of this circuit configuration are readily available for UHF applications. However, the varactors used are silicon based which behave as insulators at cryogenic temperatures due to the lack of thermal energy needed for free carrier excitation [4]. This paper describes how a commercial phase shifter was modified for cryogenic operation and presents a preliminary evaluation of its performance.

4.2 Modifications

The commercial phase shifter chosen for this evaluation was the Mini-Circuits SPHSA-152+, a voltage-variable unit housed in a surface mount package. It operates over the 800–1500 MHz range and contains two of the basic circuit blocks shown in Fig. 4.2 to achieve 360° variation. The packaged unit was mounted to a cryogenic carrier (see Fig. 4.3), designed specifically for this purpose based on many years of experience developing cryogenic low noise amplifiers [5].

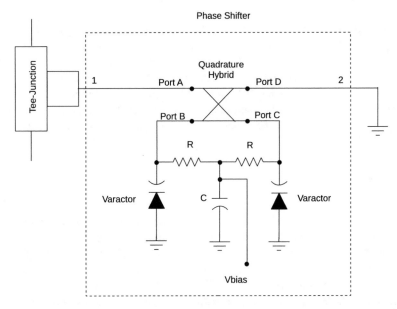

Fig. 4.2 Equivalent circuit for an electronic phase shifter [3]. DC bias voltage is used to simultaneously vary the capacitance of the varactor diodes

Fig. 4.3 Photograph of the Mini-Circuit phase shifter mounted to the cryogenic test carrier with cover removed. RF input and output ports are on the right and the DC bias connector is on the left

The phase shifter was mounted to an FR-4 circuit board which, in turn, was attached to the gold-plated brass carrier using conductive epoxy. The RF ports are 3.5 mm coaxial connectors and the bias is a 6-pin in-line connector. Bench measurements using an HP 8753D Vector Network Analyzer (VNA) confirmed that the circuit met ambient temperature specifications over the biasing voltage range.

The phase shifter was mounted to the 15 K cold plate inside the Dewar of a closed-cycle refrigeration system that has been used for cryogenic amplifier development. While carrier "freeze out" in the silicon varactors was indeed observed, the test confirmed that the remaining components of the phase shifter would hold up mechanically under the stresses of thermal cycling.

The four silicon varactors were replaced by MACOM MA46H204 Gallium Arsenide (GaAs) Hyper-abrupt varactors with a gamma = 1.25 doping profile, a junction capacitance Cj(-4 v) = 10 pF, a maximum capacitance ratio 10:1, and a reverse voltage Vr = −2 to −20 V. Proper operation of the modified circuit was confirmed on the bench using the VNA. This GaAs phase shifter was remounted in the cryogenic Dewar and cooled to approximately 15 K where a complete set of complex, two-port S-parameters were measured as a function of bias voltage.

4.3 Modeling

An equivalent circuit model, based on the configuration of Fig. 4.2 with added parasitic elements, was developed for the cryogenic phase shifter, and Keysight's Advanced Design System (ADS) was used to fit the model's component values so that the S-parameters from the model conform to the measurements at each bias setting with $\Delta S \leq 0.01$ error over the 500–1500 MHz band. The phase shift of the model versus frequency, which agrees with measurements, is shown Fig. 4.4 for bias settings of 0, −2, −4, −6, −8, and −10 V. The model was used to estimate the noise temperature of the phase shifter if it were cooled to 3 K. The results are shown in Fig. 4.5 for the same bias settings. This is an upper limit on noise since the model includes the Dewar coaxial transmission lines.

4.4 Conclusions

This early experiment has demonstrated that a commercial phase shifter can be successfully modified to operate at cryogenic temperatures. Model extrapolation to operating temperatures below 15 K reveal encouraging noise performance but further work is needed on the model to reduce the error to $\Delta S \leq 0.001$ to improve de-embedding accuracy for better identification of losses.

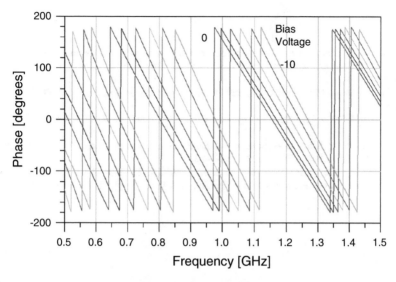

Fig. 4.4 Results of the ADS modeling for 500–1500 MHz operation at 15 K. The phase shift versus frequency at 15 K for six bias settings (0, −2, −4, −6, −8, and −10 V) is given in the plot

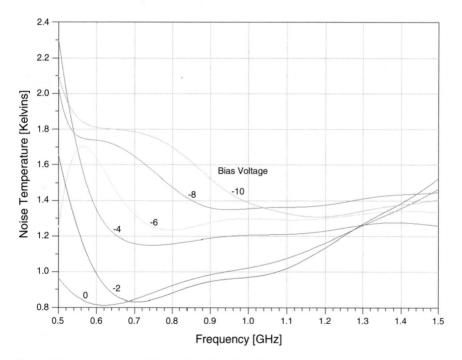

Fig. 4.5 Results from the ADS modeling for 500–1500 MHz operation at 15 K. Plot shows an estimate of the maximum noise temperature expected at a physical temperature of 3 K at bias settings of 0, −2, −4, −6, −8, and −10 V. The noise temperature is under 2.0 K over this bias range

References

1. W. Riddle, C. Nelson, Impedance control for critically coupled cavities, in *Proceedings of the 2005 IEEE International Frequency Control Symposium and Exposition* (IEEE, Piscataway, 2005)
2. R. Bradley et al., Microwave cavity searches for dark-matter axions. Rev. Mod. Phys. **75.3**, 777 (2003)
3. D.M. Pozar, *Microwave Engineering*, Addison-Wesley, Reading, MA, USA, chap. 11, p. 646 (1990)
4. K. Vinod Kumar, Extreme-temperature and harsh-environment electronics, chap. 3 (IOP, 2017). https://doi.org/10.1088/978-0-7503-1155-7ch3
5. R.F. Bradley, Cryogenic, low-noise, balanced amplifiers for the 3001200 MHz band using heterostructure field-effect transistors. Nucl. Phys. B, Proc. Suppl. **72**, 137–144 (1999)

Chapter 5
Application of the Bead Perturbation Technique to a Study of a Tunable 5 GHz Annular Cavity

Nicholas M. Rapidis

Abstract Microwave cavities for a Sikivie-type axion search are subject to several constraints. In the fabrication and operation of such cavities, often used at frequencies where the resonator is highly overmoded, it is important to be able to reliably identify several properties of the cavity. Those include identifying the symmetry of the mode of interest, confirming its form factor, and determining the frequency ranges where mode crossings with intruder levels cause unacceptable admixture, thus leading to the loss of purity of the mode of interest. A simple and powerful diagnostic for mapping out the electric field of a cavity is the bead perturbation technique. While a standard tool in accelerator physics, we have, for the first time, applied this technique to cavities used in the axion search. We report initial results from an extensive study for the initial cavity used in the HAYSTAC experiment. Two effects have been investigated: the role of rod misalignment in mode localization, and mode-mixing at avoided crossings of TM/TE modes. Future work will extend these results by incorporating precision metrology and high-fidelity simulations.

Keywords Axion · Bead-pull · Mode mixing · Mode crossing · Simulation · Perturbation technique · Dielectric · TM modes · TE modes

5.1 Background

5.1.1 Introduction

HAYSTAC is a University of California, Berkeley, University of Colorado Boulder, and Yale University experimental collaboration designed to detect the QCD axion. The experiment consists of a Sikivie type detector that detects the axion through its conversion into two photons via the Primakoff effect. One photon is virtual and

N. M. Rapidis (✉)
University of California Berkeley, Berkeley, CA, USA
e-mail: rapidis@berkeley.edu

© Springer International Publishing AG, part of Springer Nature 2018
G. Carosi et al. (eds.), *Microwave Cavities and Detectors for Axion Research*,
Springer Proceedings in Physics 211, https://doi.org/10.1007/978-3-319-92726-8_5

is provided by a strong magnetic field [1]. The ultimate range of the experiment would allow for the detection of axion masses in the 20–100 μeV range. A 9.4 T magnet provides the required magnetic field. A VeriCold Technologies dilution refrigerator maintains the physical temperature of the microwave cavity and the Josephson Parametric Amplifier at 127 mK. The microwave cavity setup consists of a 10 in. height and 4 in. diameter oxygen-free high conductivity copper cavity. A 2 in. diameter copper rod is used to tune the cavity's TM_{010}-like mode. The axis of the tuning rod is off-center from the axis of the cavity; when the rod is rotated 180°, the frequency of the TM_{010}-like mode tunes over 3.6–5.8 GHz [2].

5.1.2   Electromagnetic Properties of the Resonator

The axion conversion power in a microwave cavity is given by [2]:

$$P_{sig} \propto B_0^2 V C_{mnl} Q_L. \tag{5.1}$$

For optimal performance, it is therefore ideal to maximize the form factor (C_{mnl}) and the quality factor (Q_L). Q_L is the quality factor of a resonator which is proportional to a mode-dependent constant of order 1 and the volume and inversely proportional to the surface area and the skin depth. The form factor is defined as:

$$C_{nml} \equiv \frac{(\int d^3\mathbf{x}\, \hat{\mathbf{z}} \cdot \mathbf{e}^*_{mnl}(\mathbf{x}))^2}{V \int d^3\mathbf{x}\, \epsilon(\mathbf{x})\, |\mathbf{e}_{mnl}(\mathbf{x})|^2} \tag{5.2}$$

where V is the volume of the vacuum inside the cavity and $\epsilon(\mathbf{x})$ is the dielectric constant, which is set to 1. The TM_{010}-like mode of the cavity is used because it possesses a uniform electric field along the z-direction which gives the largest form factor of any mode.

The resonant frequencies of TM-like modes are a function of rod position whereas TE and TEM modes show virtually no dependence (Fig. 5.1). This allows for a wide range of potential axion masses to be scanned by tuning the TM modes. However, problems are introduced due to mode mixing when the frequency of the desired TM mode approaches the frequency of a TE or TEM mode.

When the frequency of the desired TM mode approaches a stationary mode (a TE or TEM mode) mode mixing occurs. The two modes in proximity hybridize, i.e. they become linear combinations of the two states as they are far apart, and thus the form factor of the initial TM-like mode diminishes by an unknown amount. The phenomenon is analogous to a two-level avoided crossing in quantum mechanics, seen in atomic and nuclear systems. This leads to regions in the frequency spectrum where the sensitivity of the experiment to axions is compromised, and thus needs to be excised from the data set. What is more problematic is that each mode crossing is different, and it is unknown over what frequency span, the data should be cut out. The bead perturbation technique allows us to examine the hybridization of the

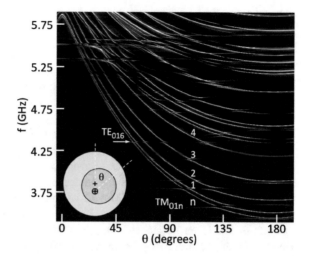

Fig. 5.1 Mode frequencies as a function of tuning rod angle. The TM modes are tuned through the TE modes whose frequencies are only weakly dependent on rod angle

modes as they approach one another, and thus guide the decision of what frequency range should be eliminated. Optimistically, it could allow us to recover some or most of the data in the mode-mixing region by measuring the TM component of the two mixed states. The results reported here are directed towards the study of such phenomena.

5.1.3 Bead Perturbation Technique

A useful tool in observing properties of certain cavity modes is the bead perturbation technique. In test cavities, a hole is made on both endcaps such that a small dielectric bead can pass through the whole length of the cavity at a constant radial and angular position. In our studies, a cylindrically shaped alumina ($\epsilon = 9.1$) bead of $h = 4.80$ mm and $r = 2.15$ mm attached to a Kevlar string traverses the cavity. As the bead travels through the cavity, the resonant frequencies of the cavity modes undergo small frequency shifts that can be calculated in perturbation theory. Specifically:

$$\frac{\Delta\omega}{\omega} = \frac{-(\epsilon - 1)}{2} \frac{V_{Bead}}{V_{Cavity}} \frac{E(\mathbf{r})^2}{\langle E(\mathbf{r})^2 \rangle_{cav}} \tag{5.3}$$

where V is the volume and ϵ is the dielectric constant of the bead [3].

This method provides a "profile" of each mode at a given rod angle by plotting the frequency of the mode as a function of bead position inside the cavity. While the frequency shift only determines the magnitude of the electric field and not its individual vector components, comparing the profile with a simulation invariably allows characteristics of modes to be determined.

5.2 Mode Mixing

Mode mixing can cause significant regions of the frequency range to become unusable due to a potentially lower Q_L and a lower form factor of the TM_{010}-like mode. To study the effect of these mode crossings, the TM_{010}-like mode was tuned to a frequency with no noticeable mixing. This was confirmed by performing a bead pull and confirming that both the TM_{010}-like mode and the intruder mode (TE or TEM mode) that were being studied, had the expected profiles. The TM_{010}-like mode was almost entirely flat whereas the intruder modes had several peaks and nodes. The TM_{010}-like mode was tuned such that its frequency increased and approached the frequency of the intruder mode. This was done by adjusting the angle of rotation of the tuning rod in small increments. As the two modes approached one another they mixed, as is evident from the field profiles in the insets of Fig. 5.2. The lower frequency TM_{010}-like mode developed an oscillatory profile, whereas the higher frequency TE mode picked up a constant offset component. In this case (though not in all such cases), the two hybrid modes never crossed; as the TM_{010}-like mode frequency continued to be changed, the lower mode remained stationary and asymptotically became the TE mode, whereas the higher mode moved upward in frequency and asymptotically became the pure TM_{010}-like mode. The mixing could also be observed by recording the Q_L of the TM_{010}-like mode through a vector network analyzer. The TM_{010}-like mode was then tuned to a

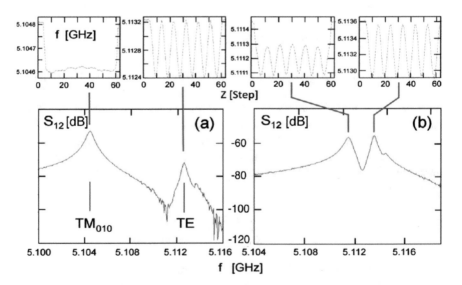

Fig. 5.2 Bead pull profiles for the frequency of the TM_{010}-like mode and a TE mode. (**a**) Corresponds to a rod position where the two peaks are not mixing. In (**b**) the lower peak (TM_{010}-like) exhibits an oscillatory behavior as a function of the position of the bead, in contrast to the case where no mixing occurs where such variations are effectively absent. The cavity extent is approximately $z = 5$ to $z = 60$ [2]

frequency higher than that of the intruder mode until it was no longer exhibiting any mixing, thus determining the range of the frequencies over which mode mixing occurs. Throughout this process, a field profile was measured for both the lower and upper mode.

While this investigation on mode crossing behavior did confirm the expected hybridization of the TM_{010}-like mode as it crossed intruder modes, a lack of mixing was observed for several intruder modes. Many modes in the frequency range of the TM_{010}-like mode were benign even though they were noticeable peaks in the spectrum.

5.3   Grid Measurements

The profile of modes as observed through the bead perturbation technique is highly sensitive to any breaking of the axial symmetry in a cylindrical cavity. If the axis of the rod is not parallel to the axis of the cavity, the E_z component of the field is no longer constant along the axis. Instead, there is localization which results in a difference in the magnitude of the frequency shift introduced by the bead at different z positions. It is therefore worth studying the effect of small misalignments in the rod's tilt.

To control misalignments, two orthogonally oriented micrometers were placed on one of the endcaps. The micrometers were positioned around the axle of the copper rod, allowing one end of the rod's axis to be translated perpendicularly to the axis of the cavity. This corresponded to a tilt in the axis of the rod with respect to the axis of the cavity. The rod was locked at a specific angle by placing a tight collar around its axle, thus inhibiting rotations.

By performing a bead pull, the resulting frequency shift due to tilting was observed. At each misalignment position, a change in frequency (Δf) was recorded which corresponded to the frequency difference of the TM_{010}-like mode with the bead at one end of the cavity compared to the other end (Fig. 5.3).

The micrometers were adjusted to take a grid of points around the central position where $\Delta f = 0$. This procedure was executed in three different tuning positions: $\theta = 0°$ corresponding to the rod being at the center of the cavity, $\theta = 90°$, and $\theta = 180°$.

Similar behavior was observed in all three tuning positions. The data formed a plane-like surface with a series of nearly flat points located on one of the diagonals (Fig. 5.4). Computing the tilt of the rod's axis with respect to the cavity's, one can calculate that each step of 0.8 mil (0.0008 in.) displacement corresponds to an angle difference of about 0.08 mrad. It is thus clear that the cavity is very sensitive to rod misalignments and hence the rod alignment needs to be carefully controlled.

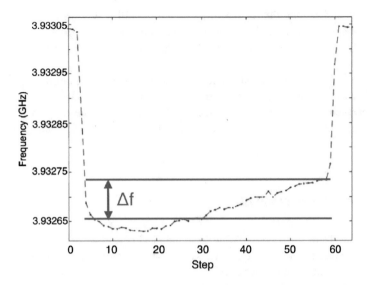

Fig. 5.3 Non-flat profile of the mode frequency for a bead pull when symmetry was broken. The cavity extent is from $z = 5$ to $z = 60$. In this case $\Delta f \equiv f(z = 5) - f(z = 60)$

Fig. 5.4 Δf as a function of micrometer displacements in the $\theta = 90°$ case

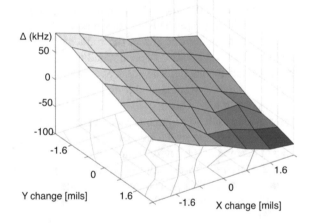

5.4   Conclusion and Future Work

A better understanding of the effects of mode crossings and misalignments in the HAYSTAC microwave cavity has been achieved leading to a better understanding of potential causes of a loss of the signal power. The effects of small rod misalignments are significant since they establish the dimensional and angular tolerances for cavity fabrication and assembly.

Future work on these aspects of the experiment will include a comparison with simulations on CST Microwave Studio. These simulations will have an accurate rendering of the rod used in this investigation created from highly precise metrology. An *in situ* bead perturbation capability is being contemplated for the actual HAYSTAC experiment for real time monitoring of the quality of the mode from which data is being measured. An open question is whether there are any misalignments induced when the cavity is cooled to <100 mK.

A forthcoming paper will report on this work and on simulations of the measurements.

Acknowledgements This work was supported under the auspices of the National Science Foundation, under grant PHY-1607417, and the Heising-Simons Foundation under grant 2014-182.

References

1. P. Sikivie, Phys. Rev. D **32**, 2988 (1985). https://doi.org/10.1103/PhysRevD.36.974
2. S. Al Kenany et al., Nucl. Inst. Meth. Phys. Res. A **854**, 11 (2017). https://doi.org/10.1016/j.nima.2017.02.012
3. J.C. Slater, Rev. Mod. Phys. **18**, 441 (1946). https://doi.org/10.1103/RevModPhys.18.441

Chapter 6
Novel Resonators for Axion Haloscopes

Ben T. McAllister, Maxim Goryachev, and Michael E. Tobar

Abstract The ARC Centre of Excellence in Engineered Quantum Systems, hosted at the University of Western Australia, has significant experience with microwave experiments, and novel resonator design. We believe there is much room for expansion and improvement in the design of resonators for haloscope searches, which are microwave cavity experiments designed to detect dark matter axions. We present schemes for novel haloscope resonators based on re-entrant cavities, dielectrics, and meta-materials.

Keywords Axion · LC resonator · Lumped element · Dielectric disk · Ring · Meta-materials · Re-entrant coils · Haloscope

6.1 Haloscope Resonant Design

Axions remain one of the most well motivated dark matter candidates, as they can be introduced as an elegant solution to the strong CP problem in QCD, and can be formulated as a significant component of the dark matter [1, 2]. The most mature experiments to detect local galactic halo dark matter axions are known as haloscopes [3], and rely on resonant conversion of axions to photons inside microwave cavities, stimulated by strong external magnetic fields. It can be shown that the haloscope scan rate is proportional to [4]

$$\frac{df}{dt} \propto \frac{1}{SNR_{goal}^2} \frac{g_{a\gamma\gamma}^4 B^4 C^2 V^2 \rho_a^2 Q_c Q_a}{m_a^2 (k_B T_n)^2}. \tag{6.1}$$

B. T. McAllister (✉) · M. E. Tobar · M. Goryachev
ARC Centre of Excellence for Engineered Quantum Systems, School of Physics,
University of Western Australia, Crawley, WA, Australia
e-mail: ben.mcallister@uwa.edu.au; michael.tobar@uwa.edu.au; maxim.goryachev@uwa.edu.au

© Springer International Publishing AG, part of Springer Nature 2018
G. Carosi et al. (eds.), *Microwave Cavities and Detectors for Axion Research*,
Springer Proceedings in Physics 211, https://doi.org/10.1007/978-3-319-92726-8_6

Here SNR_{goal} is the desired signal-to-noise ratio of the search, B is the strength of the external magnetic field, Q_a is the axion signal quality factor, Q_c is the cavity quality factor, T_n is the effective noise temperature of the first stage amplifier, with later amplifier contributions suppressed by the gain of this amplifier, V is the cavity volume, m_a, ρ_a and $g_{a\gamma\gamma}$ depend on the properties of the axion, and C is a mode dependent form factor of order 1 [5]. This scan rate is the quantity that must be maximized in design of an experiment, for which $C^2 V^2 G$ can be viewed as a figure of merit for resonator design, as these are the only resonator dependent terms, where G is the mode geometry factor, which is directly proportional to quality factor such that $Q_c = \frac{G}{R_s}$, where R_s is the surface resistance of the walls of the resonator.

Most mature haloscope searches have been focused on the frequency range of a few hundred MHz to a few GHz, but much recent work motivates axion searches in frequency ranges higher and lower than this [6–8]. However, haloscope resonator design in these high and low mass ranges is non-trivial, as a number of factors conspire to decrease feasibility and sensitivity [9]. Factors such as the size requirements for traditional resonators, amplifier availability, mode form factors, and available space inside magnet bores contribute to this problem, if one simply scales a traditional haloscope experiment to another frequency range. These problems motivate the design of new resonator schemes for such experiments.

6.2 Lumped 3D LC Resonators

The first proposed scheme is built on a structure known as a re-entrant cavity, which is a three-dimensional, lumped version of a LC resonant circuit. The cavity in question consists of a cylindrical conducting can with a central post, and a small gap between the top of the post and the lid of the cavity. As the post is inserted into the cavity the empty cavity TM modes are perturbed, and transition into re-entrant modes. The frequencies shift downwards as the post is inserted, with the fundamental TM_{010} mode frequency tending to zero as the gap tends to zero, and higher order TM modes transitioning into coaxial modes of fixed, non-zero frequency. This is beneficial for haloscopes, as it allows for lower frequencies to be attained without the need for physically large cavities, which can be challenging to implement due to size restrictions inside high field magnet bores. Figure 6.1 shows a typical re-entrant cavity, frequency tuning, and $C^2 V^2 G$ sensitivity product for such a resonator. The existence of multiple higher order re-entrant modes that are each sensitive in different regions of the tuning range allows for the possibility of searching with multiple modes within the same cavity, during the same scan. The low frequencies and immense tuning ranges attainable with these devices are also appealing for low mass axion haloscopes. See [10, 11] for more detail on the application of these resonators to axion haloscopes.

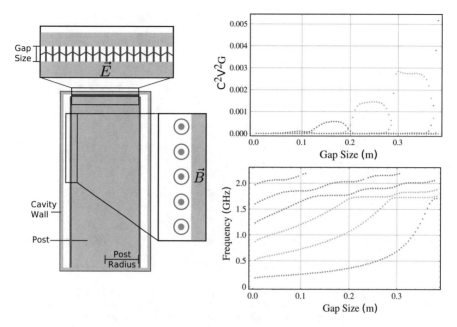

Fig. 6.1 Left: A diagram of a typical re-entrant cavity, and the mode electric and magnetic fields. Top right: C^2V^2G sensitivity product vs gap size for the first several re-entrant modes. Bottom right: resonant frequency vs gap size for the first several re-entrant modes. Blue represents the fundamental, yellow the next highest order, followed by green, red, purple and brown. Data found by finite element modelling for an arbitrarily sized re-entrant cavity

6.3 Dielectric Resonators

The use of dielectric materials in haloscopes has been of growing interest in the push towards higher mass axion searches. We present two novel schemes for dielectric haloscope resonators based on super-modes. Super-modes are pairs of modes with similar spatial structure, but where one mode has an extra variation along a given direction.

6.3.1 Dielectric Disk Resonator

The first scheme exploits axial super-modes in dielectric disk resonators. In this scheme, two dielectric disks are supported in the centre of a resonator on dielectric posts, such that the field structures shown in Fig. 6.2 are supported. These are axial super-modes, such that one mode has all magnetic field in phase (and is consequently sensitive to axions), and one has a single variation along the z-direction. As the gap size between the two disks is increased, the frequency of

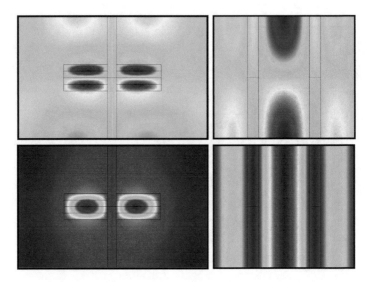

Fig. 6.2 Left: The B_ϕ field structure for the two super-modes in the disk resonator with zero gap size Right: The E_z field structure for the two super-modes in the ring resonator. In both cases the symmetric super-mode is on the bottom, whilst the anti-symmetric super-mode is on the top

the symmetric (in-phase) mode increases to approach the frequency of the anti-symmetric (out of phase) mode. The lower frequency symmetric mode retains axion sensitivity and rapidly tunes in frequency with small gap size adjustments, as per Fig. 6.3. This rapid tuning with small displacements is highly appealing for axion haloscopes. For a more detailed discussion see [4].

6.3.2 Dielectric Ring Resonator

The second scheme is similar, in that it exploits super-mode pairs in dielectric resonators. This scheme is based on hollow, dielectric cylinders placed inside resonant cavities, carefully positioned such that the out of phase field regions in higher order TM modes (TM_{020}, TM_{030}, etc.) are confined within the dielectric and suppressed by the relative dielectric constant, which leads to a boost in field uniformity, and thus a gain in form factor. Traditionally in haloscopes only the lowest order TM mode is employed, as higher order modes have significantly reduced form factors. In the push to higher frequencies, utilizing higher order modes with boosted form factors is highly promising, as it allows for physically large resonators at high frequencies, whilst retaining axion sensitivity. Furthermore, such a scheme can be tuned in a similar fashion to the dielectric disk resonator. If we split the hollow dielectric cylinder along the $r - \theta$ plane, we can exploit axial super-modes. As we increase the gap between the two sections of the ring, the sensitive,

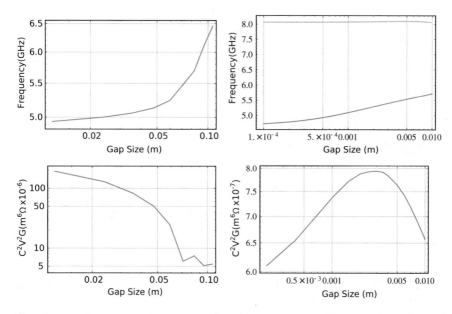

Fig. 6.3 Top Left: Resonant frequency vs gap size for the symmetric super-mode in the ring resonator. Bottom Left: C^2V^2G product vs gap size for the symmetric super-mode in the ring resonator. Top Right: Resonant frequency vs gap size for the symmetric super-mode (blue) and anti-symmetric super-mode (yellow) in the disk resonator. Bottom Right: C^2V^2G product vs gap size for the symmetric super-mode in the disk resonator. All data found via finite element modelling for structures designed to tune around 5–6 GHz

symmetric TM_{0n0}-like mode tunes rapidly upwards in frequency towards the anti-symmetric TM_{0n1}-like mode. Figures 6.2 and 6.3 show the electric field pattern, sensitivity, and tuning range for such a resonant scheme. Again, see [4] for a more detailed discussion.

6.4 Meta-materials

The final proposal relies on an array of small structures located inside a single cavity. The structures in question are "re-entrant coils", and the modes are similar in structure to the re-entrant modes discussed in Sect. 6.2. These kinds of coils are often employed in the meta-material community. The key advantage of this kind of system is scalability; if we start with a rectangular conducting box containing a single re-entrant coil the mode in which all of the electric field is in phase, which is axion-sensitive, will have a certain frequency. If we double the size of the box, and add a second identical coil we can support a mode with the same frequency and field structure as the smaller, single coil mode, but with double the mode volume. Finite element modelling predicts that this kind of system will scale in axion sensitivity

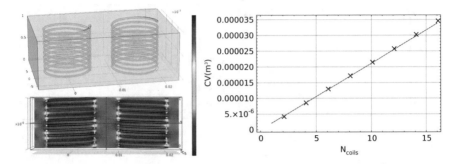

Fig. 6.4 Left: Representations of the re-entrant coils and electric field patterns of the axion-sensitive mode for arbitrarily sized coils. Right: Data from finite element modelling for a form factor and volume product as a function of the number of coils (crosses). The red line represents perfect scaling by n, the number of coils. It should be stressed that increasing the number of coils does not significantly alter frequency

indefinitely with the number of coils, and will be tunable by adjusting the gaps between the turns of the coils. Furthermore, by selective arrangement of the coils it is possible to engineer a so-called "left-handed" meta-material, with the property that the most uniform field pattern corresponds to the highest frequency mode, which means that high form factors are achieved for the highest frequency mode in the system. This is obviously very appealing in the push towards high mass axion haloscopes. Goryachev et al. [12] provides a discussion of left-handed meta-materials, and their application in axion detection (Fig. 6.4).

Acknowledgements The authors would like to thank Eugene Ivanov, Graeme Flower, Lucas Tobar, and Jeremy Bourhill.

References

1. R.D. Peccei, H.R. Quinn, Phys. Rev. Lett. **38**, 1440 (1977). https://doi.org/10.1103/PhysRevLett.38.1440
2. J. Ipser, P. Sikivie, Phys. Rev. Lett. **50**, 925 (1983). https://doi.org/10.1103/PhysRevLett.50.925
3. P. Sikivie, Phys. Rev. Lett. **51**, 1415 (1983). Erratum: [Phys. Rev. Lett. **52**, 695 (1984)]. https://doi.org/10.1103/PhysRevLett.51.1415, 10.1103/PhysRevLett.52.695.2
4. B.T. McAllister, G. Flower, L.E. Tobar, M.E. Tobar, Phys. Rev. Appl. **9**, 014028 (2018). https://doi.org/10.1103/PhysRevApplied.9.014028
5. B.T. McAllister, S.R. Parker, M.E. Tobar, Phys. Rev. Lett. **116**(16), 161804 (2016). Erratum: [Phys. Rev. Lett. **117**(15), 159901 (2016)]. https://doi.org/10.1103/PhysRevLett.117.159901, 10.1103/PhysRevLett.116.161804 [arXiv:1607.01928 [hep-ph], arXiv:1512.05547 [hep-ph]]
6. C. Beck, Phys. Rev. Lett. **111**, 231801 (2013). https://doi.org/10.1103/PhysRevLett.111.231801 [arXiv:1309.3790 [hep-ph]]
7. G. Ballesteros, J. Redondo, A. Ringwald, C. Tamarit (2016). arXiv:1610.01639 [hep-ph]

8. E. Berkowitz, M.I. Buchoff, E. Rinaldi, Phys. Rev. D **92**(3), 034507 (2015). https://doi.org/10.1103/PhysRevD.92.034507 [arXiv:1505.07455 [hep-ph]]
9. B.T. McAllister, G. Flower, E.N. Ivanov, M. Goryachev, J. Bourhill, M.E. Tobar, Phys. Dark Universe **18**, 67–72 (2017). https://doi.org/10.1016/j.dark.2017.09.010
10. B.T. McAllister, S.R. Parker, M.E. Tobar, Phys. Rev. D **94**(4), 042001 (2016). https://doi.org/10.1103/PhysRevD.94.042001 [arXiv:1605.05427 [physics.ins-det]]
11. B.T. McAllister, Y. Shen, C. Flower, S.R. Parker, M.E. Tobar, J. Appl. Phys. **122**, 144501 (2017). https://doi.org/10.1063/1.4991751
12. M. Goryachev, B.T. Mcallister, M.E. Tobar, Phys. Lett. A **382**, 2199–2204 (2018). https://doi.org/10.1016/j.physleta.2017.09.016

Chapter 7
Photonic Band Gap Cavities for a Future ADMX

Nathan Woollett and Gianpaolo Carosi

Abstract The coupling of the 'Standard Model Axion' exists within a finite band, the mass is only bounded by experiment and observation. This leads to the need for multiple experiments dedicated to searching specific mass regions. The current ADMX run plan uses cylindrical microwave cavities to act as a detector in the frequency range of 500 MHz–10 GHz. Beyond 10 GHz the relative performance of microwave cavities is reduced and therefore new technology is needed. Photonic bandgap cavities act like conventional cavities but they can operate in the 10–100 GHz with little degradation in performance. A crucial aspect of the haloscope technique employed in ADMX is the ability to tune the frequency of the cavity. In this report it will be shown that PBGs are tuneable within 10% of the fundamental frequency while retaining a high Quality factor.

Keywords Axion · Dark matter · Photonic band gap (PBG) cavities ·
Haloscope · High frequency · Defects · Symmetry breaking · Form factor ·
Tunable

7.1 Axions

The Standard Model of particle physics is a description of the universe which has stood up and adapted to many years of testing. Despite its success it is by no means complete and many observations remain unexplained. Specifically, CP violation is not observed in the strong sector despite it being allowed by current theory. One way this reveals itself is through the neutron dipole moment being measured to be over 10 orders of magnitude smaller than would be expected based on The Standard Model alone[1, 3, 5]. The axion solves this problem by treating the degree to which CP is broken as a dynamic field which through symmetry breaking provides a

N. Woollett (✉) · G. Carosi
Lawrence Livermore National Laboratory, Livermore, CA, USA
e-mail: woollett2@llnl.gov

© Springer International Publishing AG, part of Springer Nature 2018
G. Carosi et al. (eds.), *Microwave Cavities and Detectors for Axion Research*,
Springer Proceedings in Physics 211, https://doi.org/10.1007/978-3-319-92726-8_7

natural explanation for why the observed value is so low[2]. The axion particle is the pseudo-Nambu-Goldstone boson resulting from this additional field[6, 7].

The axion possesses interactions with the SM, primarily through couplings to the photon and the electron; however there have been no direct observations of the axion to date. To explain this non-observation, the rate of interactions between the axion and SM particles must be low, making axions an ideal candidate for cold dark matter. To detect axions within the μev-meV mass range, many experiments probe the axion, photon-photon coupling which is described by the Lagrangian in Eq. (7.1).

$$\mathscr{L}_{A\gamma\gamma} = \frac{G_{A\gamma\gamma}}{4} F_{\mu\nu} \tilde{F}^{\mu\nu} \phi_A, \qquad (7.1)$$

where $G_{A\gamma\gamma}$ is the coupling constant, ϕ_A is the axion field, F is the electromagnetic field strength tensor and $\tilde{F}$ is it's dual.

7.2 Haloscopes

Since axion interactions are rare, an intense source is needed to have a observable coupling rate. The galactic dark matter has an energy density of $0.3\,\mathrm{GeV\,cm^{-3}}$ making it an excellent axion source[4]. Haloscopes use a magnetic field to facilitate the axion, virtual-photon coupling and a internal resonator to allow the signal from dark matter axions to build up to an observable level. The haloscope is only sensitive to axions with an energy within the bandwidth of the resonator, because of this the resonators are tuned over a range of frequencies to cover a range of masses. The rate of tuning is described by the Dicke radiometer equation, Eq. (7.2),

$$\Delta F = \frac{t P_{sig}^2}{(k_B T \sigma)^2}, \qquad (7.2)$$

where ΔF is the detector bandwidth, t is the integration time, P_{sig} is the signal power, k_B is the Boltzmann constant, T is the noise temperature and σ is the signal to noise ratio.

At resonant frequencies of a few GHz, conventional microwave cavities make effective resonators for haloscopes. However as the resonant frequency increases, the sensitivity of a conventional cavity reduces due to the reduced volume and degradation in the RF properties of the materials.

7.3 Photonic Band Gap Cavities

Photonic structures are structures with a periodic spatial variation in the permittivity of the material. By tuning the geometric structure it is possible to control the flow of electromagnetic waves and create a bandgap in the supported frequencies. This means that it is not possible for photon state with a frequency within the bandgap to propagate through the bulk structure whereas those in the propagation band can flow freely (Fig. 7.1). By removing a scatterer and creating a defect within the structure it is possible to localize a state within the bandgap to the defect, making a structure which is analogous to a cavity.

The quality factor of a photonic bandgap (PBG) cavity is a function of the geometry and the permittivity contrast of the materials. As the frequency increases the geometry will shrink in size but leaves the quality factor unaffected; and through careful choice of materials the permittivity contrast can be maintained. This means that PBG cavities can achieve quality factors of 10^6 into the 100s of gigahertz.

A concern for the use of PBG cavities in a haloscope is the lack of an obvious tuning mechanism without breaking the geometry symmetry which is essential to produce a high Q factor. By rotating each inner scatterer around a point the 'cavity' volume can be shrunk, increasing the frequency without perturbing the geometry enough to severely effect the Q. Tuning using this method can produce a tuning range of 10% (Fig. 7.3).

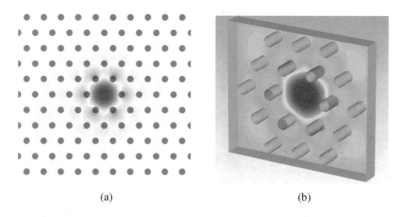

(a)   (b)

Fig. 7.1 Simulation of a defect within a photonic bandgap structure with an excited electric field of the resonant mode. The high field region is constrained to the defect with the field strength reducing exponentially away from the defect. Figure (**a**) is a 2D simulation made using the MIT MEEP Software package. Figure (**b**) is a 3D simulation made using CST Microwave Studio

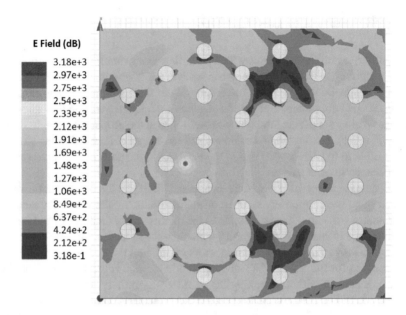

Fig. 7.2 Simulation of a photonic bandgap cavity performed using CST Microwave Studio. The defect is excited at 11 GHz which is off resonance. Since the state is not localised to the defect the energy freely flows away, reducing the number of spurious modes which can be seen within the resonator

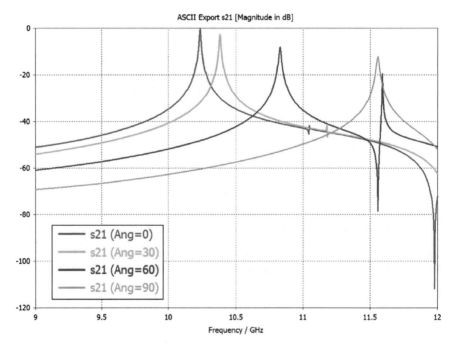

Fig. 7.3 A simulation of a S21 measurement of a photonic bandgap cavity performed with CST Mircowave Studio. The rotation angle was varied between 0–90° in steps of 30°, the resonance shifted by 1.3 GHz

Acknowledgements Lawrence Livermore National Laboratory tracking number LLNL-PROC-746080. This work was supported by the U.S. Department of Energy through Grants No. DE-SC0009723, No. DE-SC0010296, No. DE-SC0010280, No. DE-SC0010280, No. DEFG02-97ER41029, No. DE-FG02-96ER40956, No. DEAC52-07NA27344, and No. DE-C03-76SF00098. Fermilab is a U.S. Department of Energy, Office of Science, HEP User Facility. Fermilab is managed by Fermi Research Alliance, LLC (FRA), acting under Contract No. DE-AC02-07CH11359. Additional support was provided by the Heising-Simons Foundation and by the Lawrence Livermore National Laboratory and Pacific Northwest National Laboratory LDRD offices.

References

1. R.J. Crewther et al., Chiral estimate of the electric dipole moment of the neutron in quantum chromodynamics. Phys. Lett. B (1979). https://doi.org/10.1016/0370-2693(79)90128-X
2. R.D. Peccei, H.R. Quinn, CP conservation in the presence of pseudoparticles. Phys. Rev. Lett. (1977). https://doi.org/10.1103/PhysRevLett.38.1440
3. J.M. Pendlebury et al., Revised experimental upper limit on the electric dipole moment of the neutron. Phys. Rev. D (2015). https://doi.org/10.1103/PhysRevD.92.092003
4. I.J. Read, The local dark matter density. J. Phys. G: Nucl. Part. Phys. (2014). https://doi.org/10.1088/0954-3899/41/6/063101
5. A.P. Serebrov et al., New search for the neutron electric dipole moment with ultracold neutrons at ILL. Phys. Rev. C (2015). https://doi.org/10.1103/PhysRevC.92.055501
6. S. Weinberg, A new light boson? Phys. Rev. Lett. (1978). https://doi.org/10.1103/PhysRevLett.40.223
7. F. Wilczek, Problem of strong P and T invariance in the presence of instantons. Phys. Rev. Lett. (1978). https://doi.org/10.1103/PhysRevLett.40.279

Chapter 8
First Test of a Photonic Band Gap Structure for HAYSTAC

Samantha M. Lewis

Abstract Haloscopic axion searches require the tuning of a TM mode in a microwave cavity. Traditional cavities contain many unwanted modes which can result in mode crossings, ultimately reducing the effective tuning range of a cavity and slowing scan rates. Photonic band gap (PBG) structures have the potential to create resonators without TE modes, allowing for uninterrupted tuning. A tunable PBG structure has been designed for HAYSTAC. A prototype has been built and tested to validate simulations. Results of the fixed frequency case will be shown as well as details of the expected tuning.

Keywords Axion · Dark matter · Photonic band gap (PBG) cavities · Haloscope · High frequency · Defects · Symmetry breaking · Form factor · Tunable · Reciprocal · Brilliouin zone

8.1 Introduction

The axion is a hypothetical light pseudoscalar and a well-motivated candidate particle for the Cold Dark Matter (CDM). As a cousin to the π^0, it is expected to undergo an inverse Primakoff conversion in the presence of a magnetic field. This process results in a photon carrying the full energy of the axion which can then be detected. Haloscope-type detectors attempt to exploit this process to scan over possible frequency ranges in search of the axion [1]. Such detectors require quantum-limited amplifiers to reach extremely low levels of noise as well as high quality factor (Q) microwave resonators to enhance the signal. The HAYSTAC (Haloscope At Yale Sensitive To Axion CDM) detector is currently searching for axions in the 5–6 GHz range ($E \sim 20$–$25\,\mu\mathrm{eV}$) using a quantum-limited Josephson Parametric Amplifier (JPA) and a copper right cylinder microwave cavity. In order to

S. M. Lewis (✉)
University of California, Berkeley, Berkeley, CA, USA
e-mail: smlewis@berkeley.edu

© Springer International Publishing AG, part of Springer Nature 2018   67
G. Carosi et al. (eds.), *Microwave Cavities and Detectors for Axion Research*,
Springer Proceedings in Physics 211, https://doi.org/10.1007/978-3-319-92726-8_8

allow operation of the detector in higher frequency regimes, additional amplifier and cavity R&D is required. This paper focuses on current efforts to create a resonator free of mode crossings using a Photonic Band Gap (PBG) structures.

8.2 Background and Motivation

A brief overview the HAYSTAC experiment and PBG structures is provided here. For further detail, see [2–4]. HAYSTAC is a collaboration of Yale University, the University of California, Berkeley, and the University of Colorado, Boulder. Sited at Yale, the experiment uses a 9 T superconducting magnet. The bore of the magnet is 0.50 m long and 0.175 m in diameter. A dilution refrigerator cools the detector to its operational temperature of 127 mK. HAYSTAC has been collecting data since 2016 and has published results excluding the 23.55–24.0 μeV range [3]. A schematic of the detector is shown in Fig. 8.1.

8.2.1 Current HAYSTAC Cavity

The current cylindrical copper cavity uses an off-axis copper tuning rod to tune over 3.5–5.85 GHz. Tuning is achieved by rotating the tuning rod 180° from a position where the rod is tangent near the cavity wall to being concentric in the cavity barrel. These tuning positions are shown in Fig. 8.1. The loaded, critically coupled Q value of this cavity is $Q \sim 2 \times 10^4$.

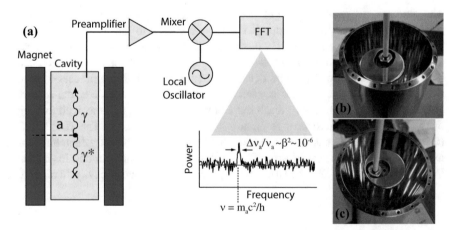

Fig. 8.1 (a) Block diagram of the HAYSTAC detector. (b) and (c) Tuning extrema for the current HAYSTAC cavity. The top photo (b) shows the minimum frequency (3.5 GHz) position and the bottom photo (c) shows the maximum frequency (5.85 GHz) position

8.2.2 Mode Crossings

Microwave cavities support resonant standing wave modes at distinct frequencies. In a typical cylindrical cavity, both transverse electric (TE) and transverse magnetic (TM) modes exist. In the current HAYSTAC cavity, the metal tuning rod gives rise to additional transverse electromagnetic (TEM) modes. Not all modes are useful in haloscope-type detectors. The detection mechanism relies on the overlap between the mode's electric field and the applied magnetic field, as described by the form factor:

$$C_{nm\ell} \equiv \frac{\left| \int d^3x\, B_0 \cdot E_{nm\ell}(x) \right|^2}{B_0^2 V \int d^3x\, \epsilon(x) \left| E_{nm\ell}(x) \right|^2}$$

where B_0 is the applied magnetic field, $E_{nm\ell}$ is the mode electric field, V is the cavity volume, and ϵ is the permittivity. The signal power is linearly dependent on the form factor and thus modes of interest are those which have the highest possible form factors. The applied magnetic field is purely in the $\hat{z}$ direction along the cavity axis, so operational modes must have a strong $\hat{z}$ component. The TM_{010} is constant in z and is thus the preferred mode of operation.

The rotation of the tuning rod changes the resonance frequencies of the TM modes but does not substantially change the frequencies of the TE or TEM modes. As the TM modes tune, they therefore encounter stationary TE and TEM modes. As one mode crosses another, the two modes mix and form a hybrid mode. The mixing reduces $C_{nm\ell}$ and the Q value, causing a reduction in signal power and rendering regions of the cavity tuning range effectively unusable. In order to scan in these regions, a different cavity barrel or tuning rod must be used. Changing the cavity requires the detector to be shut down and warmed which adds substantial downtime. There are no existing solutions to eliminate unwanted modes or reduce the impact of mode crossings.

8.3 Photonic Band Gap Structures

A photonic band gap (PBG) structure consists of a periodic lattice of metal and/or dielectric rods with a uniform size and spacing. For an infinite lattice, there exist 'band gaps' in which certain frequencies of light cannot propagate through the lattice from any direction. This behavior holds for finite lattices and has been used to create resonators and traveling-wave structures for a variety of applications [5, 6].

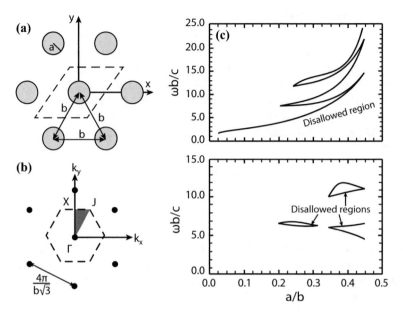

Fig. 8.2 (**a**) Example of a triangular PBG lattice, where a is the rod radius and b is the center-to-center spacing. The dashed line represents the unit cell. (**b**) Reciprocal lattice showing the Brillouin zone. (**c**) Location of band gaps for TM modes (top) and TE modes (bottom) in a triangular lattice of rods. The 'disallowed' regions are the band gaps. Modes within this region are confined when a defect is created (adapted from [7])

8.3.1 Resonators

In a PBG resonator, a subset of rods is removed to create a defect. Modes can be excited in the defect in the same manner as in a traditional cavity. However, modes which fall outside the band gaps propagate out of the lattice and are not properly confined. Modes within the band gaps cannot propagate through the lattice and are therefore confined. Figure 8.2 shows an example PBG lattice and resulting band gaps.

8.3.2 Application to HAYSTAC

A PBG resonator used in the place of a traditional microwave cavity in HAYSTAC would eliminate mode crossings by only confining TM modes. However, the structure must be tunable which is not an inherent feature of a PBG lattice. Breaking symmetry could result in the reintroduction of unwanted modes or low quality factor values, while symmetry-preserving tuning mechanisms could be prohibitively

complicated. The prototype design discussed in this work examines one possible
tuning mechanism.

Any useful HAYSTAC resonator also requires a good form factor, a good Q, and
a reasonably large tuning range. If the TM_{010}-like mode is used, the form factor
in a PBG resonator would remain high. PBG structures can have very high Q
values, however obtaining good electrical contact can be nontrivial. Tuning range
is dependent on the tuning mechanism and must be a key design goal.

8.4   Prototype Design

A prototype PBG resonator has been designed and fabricated to serve as a proof-
of-concept and a first test of a tunable PBG structure. This prototype is designed
to fit within the existing HAYSTAC magnet and thermal shield system. It has been
designed to tune in the 7.5–9.5 GHz range, although the design concepts used can
be scaled to other frequencies. The length of the structure was chosen to be 10 cm
for ease of fabrication. The TM mode frequencies are not affected by the length of
the structure, allowing the length to be scaled up as needed.

8.4.1   Lattice

The prototype lattice was designed to be easily fabricated and to interface with
existing test setups for current HAYSTAC cavities. It consists of 3.175 mm (quarter
inch) aluminum rods spaced to achieve $a/b = 0.43$. Aluminum top and bottom
end caps hold the rods in place with recessed pockets for each rod. The six corner
rods are tapped in order to screw the end caps in place. Photos of the lattice and the
simulated TM_{010}-like mode are shown in Fig. 8.3.

8.4.2   Tuning Mechanism

As a simple first test, this PBG resonator was designed to use the same tuning
mechanism as the existing HAYSTAC cavity: an off-axis tuning rod. The rod was
designed to have radius $r = 1.1b$. It was fabricated from aluminum and has a quarter
inch alumina axle in order to be easily integrated with existing setups. A photo of
the tuning rod in the structure is shown in Fig. 8.3. This tuning rod provides roughly
2 GHz of tuning range between 7.5–9.5 GHz.

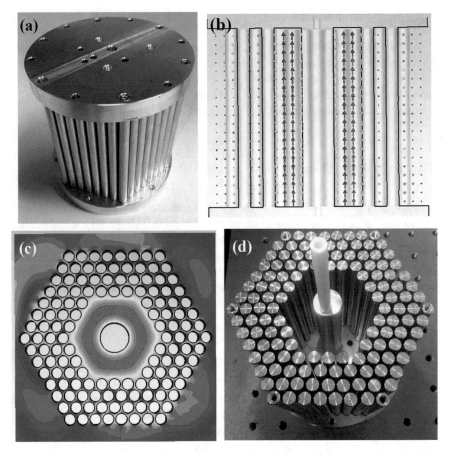

Fig. 8.3 (**a**) Full lattice assembled without tuning rod. (**b**) Simulated electric field of the TM_{010}-like mode with the tuning rod at its maximum frequency position. (**c**) Top-down view of the simulated confinement. (**d**) Partially-assembled structure with tuning rod

8.4.3   First Tests

Fixed frequency tests show excellent agreement with simulations. There have been no TE modes observed in the fixed frequency case and initial tests with the tuning rod do not show any unexpected modes. The TM mode Q values are lower than predicted, likely an issue of electrical contact between the rods and end caps. The measured tuning range of the rod matches the range predicted from simulations. Detailed results with be reported in a forthcoming paper.

8.5   Discussion and Future Work

This prototype design has demonstrated it is possible to tune a PBG structure in a symmetry-breaking manner without destroying the band gap behavior. No TE modes were introduced by the tuning rod and TM modes can tune over a wide range. However, performance is currently limited by low Q values. It is possible that more force is needed to achieve good electrical contact between the rods and the end caps. Future tests will determine the source of the discrepancy and attempt to improve the electrical contact.

Acknowledgements This work was supported under the auspices of the National Science Foundation, under grant PHY-1306729, and the Heising-Simons Foundation under grant 2014-182.

References

1. P. Sikivie, Phys. Rev. Lett. **51**, 1415 (1983)
2. T.M. Shokair et al., Int. J. Mod. Phys. A **29**, 1443004 (2014)
3. B. Brubaker et al., Phys. Rev. Lett. **118**, 061302 (2017)
4. S.A. Kenany et al., Nucl. Instrum. Methods A **854**, 11 (2017)
5. E.A. Nanni et al., Phys. Rev. Lett. **111**, 235101 (2013)
6. E.I. Smirnova et al., Phys. Rev. Lett. **95**, 074801 (2005)
7. E.I. Smirnova et al., J. Appl. Phys. **91**, 960 (2002)

Chapter 9
Hybrid Cavities for Axion Detectors

Ian Stern and D. B. Tanner

Abstract Hybrid cavities, cavities with a layer of superconductor applied on top of the OFHC copper, are studied. The quality factor of a single thin layer and of a multilayer superconductor/insulator stack are estimated. The results for these thin-film coatings are not good. However, if the superconductor (either multilayer or single layer) were separated from the copper by a thick dielectric spacer, one that is a fraction of the wavelength at the cavity resonant frequency, extremely high quality factors could be obtained.

Keywords Axion · Dark matter · Haloscope · Superconducting · Thin film · Multilayer · Anomalous skin depth · Quality factor · Type-II · Fluxoids · London model · Drude model

9.1 Introduction

The ADMX cavities are limited by the anomalous skin depth in the copper walls to unloaded quality (Q) factors of about 350,000 (at 1 GHz). It has been suggested [1] that a thin coating of superconductor would improve this value. The idea was based on work by Xi et al., [2] where it was shown that a thin film superconductor could have good superconducting properties in a magnetic field *so long as the film was thinner than the spacing of fluxoids in the type-II superconductor* and if the field were applied parallel to the sample surface. The film thickness for 8 T fields would be about 100 Å. In this regime, the optical conductivity, shown in Figure 2 of [2], has a dissipative part (σ_1) that is zero (within error) below the superconducting gap whereas the reactive part (σ_2) rises as $1/\omega$ making the loss tangent smaller and smaller as frequency decreases. The superconducting gap is reduced by field-induced pairbreaking but even at 10 T it about 15 cm^{-1} (450 GHz), well above the

I. Stern · D. B. Tanner (✉)
Department of Physics, University of Florida, Gainesville, FL, USA
e-mail: tanner@phys.ufl.edu

© Springer International Publishing AG, part of Springer Nature 2018   75
G. Carosi et al. (eds.), *Microwave Cavities and Detectors for Axion Research*,
Springer Proceedings in Physics 211, https://doi.org/10.1007/978-3-319-92726-8_9

ADMX operating frequency. Note that films in the 100 Å thickness range have finite transmission; indeed it was by reflection and transmission measurements that the data of [2] were obtained.

The lossless behavior only occurs when the static magnetic field is parallel to the surface and the film is thin. Once flux vortices appear in significant number, the film becomes lossy, comparable to the loss in the normal-state of the metal [3]. By 6–10 T the gap is filled in and σ_2 is quenched. Moreover, the normal state conductivity of type II superconductors, such as NbTi, is hundreds of times smaller than that of clean metals such as Cu; their quality factors are correspondingly small.

Although the idea of such a hybrid cavity seems promising, there has not been (to our knowledge) any actual modeling of the effect of the superconducting layer on the Q of a cavity. So this paper is a first attempt at such a model.

9.2 Microwave Properties of Superconductors

To start, because 1 GHz $\ll$ 450 GHz, we can use a London model for the superconducting electrodynamics. This model produces the Meissner effect and the infinite dc conductivity. The optical conductivity for $\omega > 0$ is purely imaginary:

$$\sigma_s(\omega) = \sigma_1 + i\sigma_2 = \frac{\pi n_s e^2}{2m}\delta(\omega) + i\frac{n_s e^2}{m\omega}. \tag{9.1}$$

The dc current is carried by the $A\delta(\omega)$ delta function at the origin. The quantity n_s is the superfluid density; it is measured through the superconducting plasma frequency $\omega_{ps} = \sqrt{4\pi n_s e^2/m}$. The other quantities in this equation are the electronic charge e and mass m. The measurements of Xi et al. [2] give $\omega_{ps} = 7.5 \times 10^{14}\,\text{s}^{-1}$. The only other thing we need is the normal-state conductivity of NbTi. We take this to be $\sigma_{dc} = 8500\,\Omega\,\text{cm}^{-1}$, given by a Drude model, $\sigma_{dc} = \omega_p^2\tau/4\pi$ with conduction electron plasma frequency $\omega_p = 6.0 \times 10^{15}\,\text{s}^{-1}$ and electron scattering rate $1/\tau = 3.8 \times 10^{14}\,\text{s}^{-1}$.

These parameters give an optical conductivity and transmittance of a thin film that reasonably resemble the data of [2]. We will also need a model for copper which will lie under the superconductor. To be correct, we should use the anomalous skin effect but the mix of local and non-local electrodynamics would require analysis beyond the scope of this work. So we use a Drude model with a smaller number of effective carriers than copper and which gives at 1 GHz identical ordinary and anomalous skin depths. The results will not have the correct frequency dependence but will illustrate the physics.

9.3 Calculations of Q

Next, we will calculate the Q of hybrid cavities. We know how to do this calculation for a metal wall, following Jackson [4]. The general expression for the unloaded Q is

$$Q = (\text{Geometric factor})\frac{V}{S\delta} \qquad (9.2)$$

where V is the cavity volume, S the cavity surface area, and δ the RF penetration depth. The (Geometric factor) depends on the mode; for the TM_{010} mode in a right circular cylinder it is equal to 2. Equation (9.2) is quite reasonable: the energy stored in the cavity is proportional to the volume of the cavity; the energy lost per cycle is proportional to the volume of the skin. After specializing to a cylinder, Eq. (9.2) becomes

$$Q = \frac{H}{R+H} \cdot \frac{R}{\delta}, \qquad (9.3)$$

where H is cavity height and R cavity radius.

The factor R/δ occurs for all modes, not just the TM_{010} mode. However, R determines the resonant frequency. The TM_{010} resonant frequency is

$$f_0 = \frac{c}{2.61R} \qquad (9.4)$$

or $R = c/2.61 f_0 = \lambda_0/2.61$ where λ_0 is the vacuum wavelength corresponding to the resonant frequency. Hence, we can use $R/\delta = \lambda_0/2.61\delta$ in Eq. (9.3) and wherever else it appears.

We can separate the losses in the sidewalls of the cylinder and the losses in the two endcaps because they add in parallel, so that $1/Q = 1/Q_s + 1/Q_e$. Writing the surface area $S = S_s + S_e$ with $S_s = 2\pi RH$ and $S_e = 2\pi R^2$, and using $V = \pi R^2 H$ as well as Eq. (9.4), we get

$$Q_s = \frac{\lambda_0}{2.61\delta} \qquad (9.5)$$

for the sidewalls and

$$Q_e = \frac{H}{R}\frac{\lambda_0}{2.61\delta} \qquad (9.6)$$

for the endcaps. It is interesting that the sidewall quality factor, Q_s, does not depend on the height of the sidewalls. The reason is that both the cavity volume and the sidewall surface area increase linearly with that height, and so their ratio does not change.

We are interested in applying the coating to the sidewalls of the cylinder, to which the external magnetic field is parallel. So for now, we consider the quality factor due to the sidewalls alone. We will add the endcaps in parallel at the end.

Now, recall the reflectance of a metal in the long-wavelength limit:

$$\mathcal{R} = 1 - \sqrt{\frac{8\omega}{\omega_p^2 \tau}} = 1 - \sqrt{\frac{2\omega}{\pi \sigma_{dc}}}. \tag{9.7}$$

This equation is called the Hagen-Rubens reflectance. The absorbance is

$$\mathcal{A} = 1 - \mathcal{R} = \sqrt{\frac{2\omega}{\pi \sigma_{dc}}} \tag{9.8}$$

because the transmittance is zero. Now, the skin depth in the classical limit is

$$\delta = \sqrt{\frac{c^2}{2\pi \omega \sigma_{dc}}} = \sqrt{\frac{2c^2}{\omega_p^2 \omega \tau}}. \tag{9.9}$$

After more algebra, we arrive at the neat equation:

$$\mathcal{A} = \frac{4\pi \delta}{\lambda_0}. \tag{9.10}$$

We can now use Eq. (9.10) to recast Q_s in terms of the absorbance of the sidewall, something that we can calculate even if there are multilayer coatings on the wall:

$$Q_s = \frac{4.81}{\mathcal{A}}. \tag{9.11}$$

We will use Eq. (9.11) not only for the copper cavity but for the coated cavity. From the point of view of the loss of the stored energy, there is no difference between a homogeneous sidewall and a layered sidewall. We can calculate the reflectance (and absorbance) of a multilayer film on a thick substrate using the ABCD matrix method described in Heavens [5].[1] Then we convert this to Q using Eq. (9.11).

---

[1] The thin-film program ThinFilmFit or TFF is available at http://www.phys.ufl.edu/~tanner/datan. html as part of the zip file datan.zip. A manual can be found at the same link. Users may probably also want to employ CoMPute (CMP) and Dielectric Function Calculator (DFC) in addition.

9.4   Results for the Sidewall Q_s for Superconductor on Copper

Figure 9.1 shows the sidewall Q_s versus frequency from just above 100 MHz to 10 GHz. The curves are (bottom to top) a NbTi cavity in its normal state, a copper cavity (at low temperatures), 100 Å of superconducting NbTi on Al$_2$O$_3$ on copper, and (topmost) a multilayer stack to be discussed shortly. Note that we know that an insulating spacing is necessary to avoid proximity-effect destruction of the superconductivity in NbTi. In the model the presence or absence of this insulator has absolutely no effect on Q_s.

The superconductor does increase Q_s but by a rather insignificant amount. It does not do what we had hoped. On the good side, a thin layer with lower conductivity has no detrimental effect.

So what to do? We have already faced the necessity of depositing two layers. How about more? We plot as the topmost curve in Fig. 9.1 the Q_s factor with 10 layers of NbTi, alternating with 10 layers of Al$_2$O$_3$. All layers are 100 Å in thickness.

The multilayer coating does give a significant increase in Q_s, although not as much as one would hope, especially as frequency increases. The increase of frequency increases the loss in the copper while at the same time the screening effect of the $1/\omega$ behavior of σ_2 is diminished. An increase of $1.5\times$ around 1 GHz

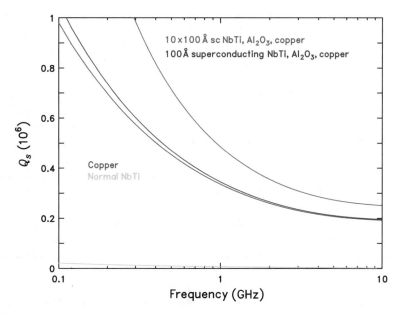

Fig. 9.1 Sidewall quality factor for a cavity made with a multilayer stack consisting of 100 Å superconducting NbTi layers separated by thin insulating layers and backed by copper. Curves for a single NbTi layer backed by insulator and copper, for high purity copper at low temperatures, and for normal-state NbTi are also shown

could be envisioned. If the superconducting layers were to become normal, the Q_s is slightly reduced (by 0.1% or less) compared to pure copper.

9.5   Sidewall Q_s with Thick Spacer Between Superconductor and Copper

Multilayer films of two materials are commonly used in high-performance coatings for laser reflectors and 20–24 layers are not uncommon. It is also true that for the superconductor/insulator coatings, the requirement on exactness of layer thickness would be decreased compared to the optical coatings.

Such multilayer optical coatings use interference effects to achieve high reflectance: the phase gain by electromagnetic waves on one round trip in each layer is set (by fixing the thickness) so that all the partial waves in reflectance add constructively. This basically means that the thickness of a layer is $\lambda_0/4n$ where $n = \sqrt{\epsilon}$ is the refractive index. (This statement is exact for zero-loss dielectric coatings, where the phase shift on reflection is neither 180° nor 0°, depending on whether the media are relatively more or less optically dense. It is incorrect for metals, where phase shifts are determined by the relative importance of the real and imaginary parts of the refractive index.) So we have taken a look at the effect of insulator spacers whose thicknesses are comparable to the wavelength.

Figure 9.2 shows the Q_s of hybrid cavities for 5 configurations, along with the Q of a copper cavity. The curves (from top to bottom) represent (1) a 5 layer structure consisting of superconducting NbTi (100 Å), 1 cm of Al_2O_3, superconducting NbTi (100 Å), 1 cm of Al_2O_3, and copper; (2) a 5 layer structure consisting of superconducting NbTi (100 Å), 1 mm of Al_2O_3, superconducting NbTi (100 Å), 1 mm of Al_2O_3, and copper; (3) a 3 layer structure consisting of superconducting NbTi (100 Å), 1 cm of Al_2O_3, and copper; (4) a 3 layer structure consisting of superconducting NbTi (100 Å), 1 mm of Al_2O_3, and copper; (5) 20 layers alternating between NbTi (100 Å) and Al_2O_3 (100 Å) and copper. The 20 layer calculation and the copper alone are also shown in Fig. 9.1. Note that the maximum Q_s calculated for this superconducting structure is $\sim 10^{13}$. Note also that the cavity top and bottom are neglected; the Q_s is the contribution of the cylinder sidewalls alone. Any tuning rod would be coated in the same way.

Experimentally, bulk superconducting cavities typically deliver $Q \sim 10^{10}$ due to residual losses in the superconductor. This loss could be accommodated in the model by including a small density of normal electrons (quasiparticles) in the dielectric function of the superconductor.

With a spacer thickness comparable to the wavelength, the spacer forms a Fabry-Perot resonator. The low transmittance of the superconductor, with corresponding reflectance approaching unity, makes this resonator have a very high finesse. The resonator is highly reflective except for the case where the spacer thickness is $\lambda_0/2n$. In this case, the light leaking out of the cavity interferes destructively with the light

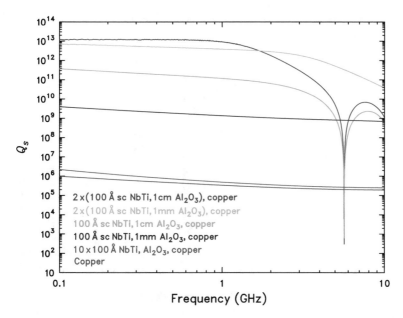

Fig. 9.2 Sidewall quality factor for cavities made of two or one NbTi superconducting layers followed by 1 cm or 1 mm of Al$_2$O$_3$ and by copper. Curves for a multilayer stack consisting of 100 Å superconducting NbTi layer separated by thin insulating layers and backed by copper and for high purity copper at low temperatures are also shown

promptly reflected from the front surface of the resonator and there is a deep and sharp minimum in the reflectance. The minimum occurs near 5.6 GHz for the 1 cm spacer and at 56 GHz (off scale) for the 1 mm spacer.

Calculations of the sidewall Q_s for three-layer (one superconducting film) and five-layer (two superconducting films) coating give results above those of bulk superconducting cavities. Hence, we now turn our attention to the sidewall Q_s of cavities with a single superconducting film, dielectric layer, and thick copper layer. Conceptually, this part of the cavity would be made by taking a dielectric tube, made of, for example, alumina or sapphire and coating the inside with 100 Å of NbTi and the outside with a thick (compared to the skin depth) copper layer. Figure 9.3 shows the calculated wall Q_s for a NbTi–Al$_2$O$_3$ spacer–copper hybrid cavity as a function of frequency for a set of spacer thicknesses in the 100 Å to 10 cm range. Thicker spacers yield higher Q_s-factors at the lowest frequencies. These thick layers have periodic minima in the Q_s over the frequency range shown. The first minimum for the thinner films is not shown; it occurs at higher frequencies.

All of the simulations with thick Al$_2$O$_3$ spacers provide Q_s values between 10^9 and 10^{12} over 300 MHz–3 GHz, seemingly more than adequate. Tuning rods would have an inverted structure: a single NbTi film on the outside of a mm-thick alumina sleeve that slips over the copper rod.

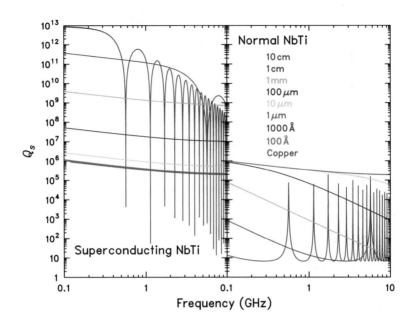

Fig. 9.3 Left panel: Sidewall quality factor for cavities made of one NbTi superconducting layer followed by an Al_2O_3 spacer and then by copper. Curves for spacer thicknesses between 100 Å to 10 cm are shown. Right panel: Sidewall quality factor for cavities made of one NbTi *normal-state* layer followed by an Al_2O_3 spacer and then by copper

The electrical currents in the TM_{010} mode flow up (or, a half cycle later, down) the cylinder walls and then out onto the endcaps. (The currents induce the $\hat{\phi}$-oriented RF magnetic field and the currents charge up the endcaps; the charges are the sources for the $\hat{z}$-oriented RF electric field.) If superconducting films are applied to the walls, a low impedance electrical circuit must be available to lead the current from the superconducting film on the alumina tubes to the copper endcaps. How low? A good estimate would be on the order of $0.5\,\mu\Omega$ if the Q_s is to be 10^8.

With high payoffs come high risks. Increasing the spacer thickness makes the Q_s *worse* if the films become normal This effect is illustrated in the right panel of Fig. 9.3. As already discussed, the normal layer has little effect when thin spacers are used. The reason (we think) is that the electric field goes to zero at the surface of the highly conducting copper and so is small in the rather resistive NbTi normal metal. When the NbTi layer is spaced further away from the copper, the fields, currents, and dissipation all increase. Near the λ_0 thickness the Q_s becomes about 10, compared to 10^5 for the copper. It is interesting that at the exact $\lambda_0/2$ point the Q_s returns to that of copper. Note that for the parameters used for NbTi, a resistivity of $\sim\!120\,\mu\Omega\,cm$ (microOhmcm) and at thickness of 100 Å, the sheet impedance is $R_\square = 120\,\Omega$, about ideal for maximizing absorption in the film as a free-standing layer.

9.6 Q Including the Endcaps

Up to now, we have calculated only the contributions to Q from the sidewalls. Tuning rods can be coated with the superconductor and would therefor add less than they do in the present cavities. Now we will add the endcaps to find the "unloaded Q" of the hybrid cavity, using $1/Q = \sum 1/Q_i$.

First, measurements shown in [3] indicate why we cannot coat the endcaps with superconductor, even superconductor spaced by a dielectric. Here, the magnetic field is perpendicular to the superconducting thin film and penetrates as flux vortices. With increasing field, these vortices appear in significant number and the film becomes lossy, comparable to the normal-state of the metal. By 6–10 T the gap is filled in and σ_2 is reduced; the contribution to the Q by the endcaps would be the same as caps made from high-resistivity, normal-state NbTi.

To estimate the contribution of the endcaps, we return to the general equation for the Q of a cylindrical cavity [4]:

$$Q = \frac{V}{S\delta} \text{(Geometrical Factor)}, \qquad (9.12)$$

where V is the cavity volume, S the cavity surface area, δ the skin depth, and the Geometrical Factor is a number of order unity, whose specific value depends on the mode, the type of cylinder, and the cavity length to radius ratio. For the TM_{010} mode of a right circular cylinder, Geometrical Factor $= 2$. The volume of the cylinder is $V = \pi R^2 H$. We assume that the walls have negligible loss, so that the surface area is only that of the 2 endcaps, making $S = 2\pi R^2$. Then

$$Q_e = \frac{H}{\delta}. \qquad (9.13)$$

We employ Eq. (9.10) to eliminate the skin depth in favor of the absorptivity $\mathscr{A}$, Eq. (9.4) to eliminate the wavelength, and the typical length to radius ratio used by ADMX cavities of $H = 5R$ to get

$$Q_e = \frac{24}{\mathscr{A}}. \qquad (9.14)$$

The anomalous skin effect formulae can be used for the endcaps. In this case $\delta \sim f^{-1/3}$ and $\mathscr{A} \sim f^{2/3}$. Combining all of the various factors, we find

$$Q_e = 1.98 \times 10^6 \left(\frac{f}{1\,\text{GHz}}\right)^{-2/3}. \qquad (9.15)$$

Figure 9.4 shows the Q for a hybrid cavity. The walls have one NbTi superconducting layer followed by a 1 mm thick Al_2O_3 spacer and then by copper. The endcaps are copper in the anomalous skin-effect limit. The endcaps dominate the

Fig. 9.4 Quality factor for a
hybrid cavity. The Q for the
walls, for the endcaps, and
the total Q are shown, along
with the Q for an all-copper
cavity. The walls have one
NbTi superconducting layer
followed by a 1 mm thick
Al_2O_3 spacer and then by
copper. The endcaps
dominate the total

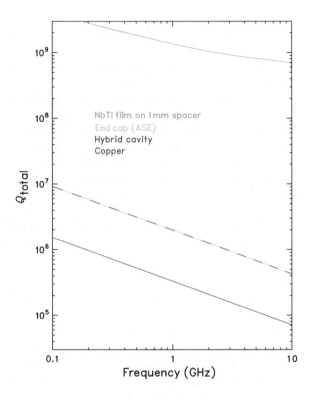

total. The hybrid cavity has a Q that is about 6 time larger than an all-copper cavity.
This gain would increase both the signal and the scan rate from an axion detector
tenfold. The gain would be achieved even if the wall Q were reduced by a factor of
10 or even 20.

9.7   Conclusions

Finally, we can comment on the requirements to produce a usable hybrid cavity.
First, the superconducting thin film must have an upper critical field higher than the
operating field of the detector. This makes NbTi or NbTiN the likely choices. A-15
superconductors, such as V_3Si and Nb_3Sn might be used, although there is evidence
[6] for gap anisotropy in these materials that might lead to greater absorbance than
NbTi. Second, the film must be thin, less than the spacing between vortices in the
superconductor. The areal density n of vortices is $n = B/\Phi_0$, where B is the applied
field and Φ_0 is the flux quantum. The average spacing is of order $a = \sqrt{1/n}$ or
160 Å for an 8 T field.

Third, the magnet must be designed so that the field lines are parallel to the cavity
walls. The allowable amount of flux penetration due to perpendicular components of

the field is hard to estimate, but one can imagine that the density must be thousands of times smaller than the 10^{15} flux quanta per m^2 of a perpendicular field.

Fourth, the superconductor must be coated on the inside of a mm-thick insulating cylinder that fits inside thick copper or copper-plated cavity walls. There must be a method to ensure that the normal-metal endcaps make good electrical contact to the copper outer wall and with the superconducting coatings.

Acknowledgements This research was supported by DOE grant DE-SC0010296.

References

1. T.M. Shokair, J. Root, K.A. Van Bibber, B. Brubaker, Y.V. Gurevich, S.B. Cahn, S.K. Lamoreaux, M.A. Anil, K.W. Lehnert, B.K. Mitchell, A. Reed, G. Carosi, Future directions in the microwave cavity search for dark matter axions. Int. J. Mod. Phys. A **29**, 1443004 (2014)
2. X. Xi, J. Hwang, C. Martin, D.B. Tanner, G.L. Carr, Far-infrared conductivity measurements of pair breaking in superconducting $Nb_{0.5}Ti_{0.5}N$ thin-films induced by an external magnetic field. Phys. Rev. Lett. **105**, 257006/14 (2010)
3. X. Xi, J.-H. Park, D. Graf, G.L. Carr, D.B. Tanner, Infrared vortex-state electrodynamics in type-II superconducting thin films. Phys. Rev. B **87**, 184503/15 (2013)
4. J.D. Jackson, *Classical Electrodynamics* (Wiley, New York, 1975)
5. O.S. Heavens *Optical Properties of Thin Solid Films* (Dover, New York, 1955)
6. D.B. Tanner, A.J. Sievers, Far-infrared measurements of the energy gap of V_3Si. Phys. Rev. B **8**, 1978–1981 (1973)

Chapter 10
An Introduction to Superconducting Qubits and Circuit Quantum Electrodynamics

Nicholas Materise

Abstract Superconducting qubits have matured from platforms demonstrating and manipulating macroscopic coherent quantum states to realizing exotic quantum states, running surface error correction codes, and single photon detection to name a few recent milestones. This article will review the fundamentals of circuit QED related to the design and simulation of superconducting qubits.

Keywords Axion · Dark matter · Qubits · Simulations · Comsol · Modeling · Numerical · Cavity · Josephson junction · Circuit model

10.1 Introduction

Superconducting qubits and circuit quantum electrodynamics have enabled design of solid state sources of quantum information. The performance of these devices has scaled exponentially over the last 15 years, in terms of their energy relaxation and dephasing times, drawing interest from adjacent communities including the Axion Dark Matter Experiment (ADMX). Recently, superconducting qubits have been suggested to be used as signal-photon detectors which could greatly increase scan rates of axion haloscopes, such as ADMX, at higher frequencies. The goal of this article is to give members of the ADMX community an introduction to some of the models used to analyze and design superconducting qubits. This review is not an exhaustive coverage of the field, but it aims to guide the reader to relevant literature and analysis techniques that closely follow experiment.

N. Materise (✉)
Lawrence Livermore National Laboratory, Livermore, CA, USA
e-mail: materise1@llnl.gov

© Springer International Publishing AG, part of Springer Nature 2018
G. Carosi et al. (eds.), *Microwave Cavities and Detectors for Axion Research*,
Springer Proceedings in Physics 211, https://doi.org/10.1007/978-3-319-92726-8_10

10.2 Superconducting Qubit Circuit Models

A qubit is a two level system or a system whose controllable quantum dynamics involve its two lowest lying energy levels. Nature provides several forms of qubits or carriers of quantum information including single photons, trapped ions, and atoms in high finesse cavities. Superconducting qubits realize *artificial atoms* with engineered energy levels using the non-linearity of Josephson junctions and surrounding microwave circuitry [2]. The quantum dynamics of these systems follows that of a damped and driven anharmonic oscillator whose anharmonicity is controlled by choice of circuit parameters, e.g. linear capacitance and inductance of the Josephson junction [14]. For experimental design and control, practitioners draw from the Jaynes-Cummings model and its variants from cavity quantum electrodynamics (QED) [8, 14]. Circuit quantum electrodynamics borrows the application of second quantized Hamiltonians from atomic optics via a standardized procedure for quantizing passive circuit. This section will introduce simple models for Josephson junctions and their role in superconducting qubits. We will then discuss circuit quantization methods and Black box quantization techniques used to obtain second quantized Hamiltonians.

10.2.1 Non-linearity in Superconducting Qubits

The operational modes of superconducting qubits vary by their energy spectra, where non-linearity plays a role in realizing accessible and isolated states. If we consider the lowest two levels of the quantum harmonic oscillator to be the ground and excited states of a qubit ($|g\rangle$, $|e\rangle$), the energies for the two states are separated by integer multiples of $\hbar\omega$. The classical electric circuit model for an oscillator is the LC circuit, shown in Fig. 10.1. We will refer to this model in Sect. 10.2.3 when we derive the second quantized form of the Hamiltonian for an LC circuit. Figure 10.1 compares the LC oscillator circuit to an anharmonic qubit, the transmon. Notice that the spacing between the excited state $|e\rangle$ and the next highest state $|f\rangle$ is smaller than the spacing between $|g\rangle$ and $|e\rangle$. In more anharmonic oscillators, the spacing is larger, further isolating the qubit states from the other states of the oscillator. The transmon trades off its anharmonicity for reduced sensitivity to charge noise [2].

An anharmonic oscillator-based superconducting qubit inherits its non-linearity from Josephson junctions, where the non-linearity is tunable through fabrication and microwave circuit design. To develop some intuition for the dynamics of Josephson junctions, we will discuss classical circuit models for the device and their role in superconducting qubits.

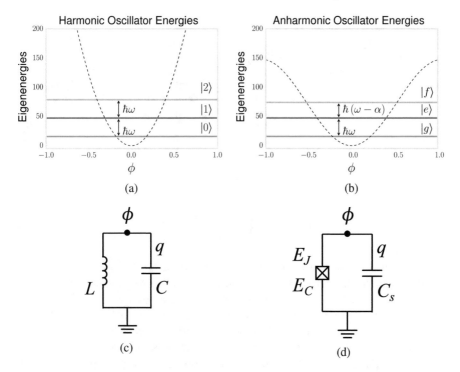

Fig. 10.1 Comparison of the quantum harmonic oscillator with an harmonic oscillator. (**a**) and (**b**) give the eigenenergies of the two oscillators, where the horizontal lines are the eigenenergies and the dashed lines represent notional potentials. (**c**) and (**d**) are the corresponding circuit models for an LC circuit and a transmon qubit [10]

10.2.2   Classical Circuit Models of Josephson Junctions

There are several phenomenological models for Josephson junctions that are motivated by the underlying device physics and limits of the electric circuit analogs. We will review the Resistive and Capacitively Shunted Junction (RCSJ) model as outlined in [5].

In Fig. 10.2 above, the left most circuit shows a current-driven Josephson junction with drive current, I_d. The junction is approximated as the parallel combination of an inductor L_J, conductance G_N, and capacitor C_J. We replace the inductor and conductance with two voltage controlled current sources (VCCS's), $G_J(\varphi), G_N(V)$, where we use the Gxxx VCCS notation from SPICE [11]. Kirchhoff's current law at the node joining the three circuit elements with the drive current source reads [5]

$$I_d(t) = I_c \sin \varphi + V G_N(V) + C_J \frac{dV}{dt} \qquad (10.1)$$

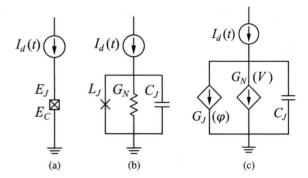

Fig. 10.2 (a) Circuit diagram for a current driven Josephson junction, (b) RCSJ circuit model, (c) equivalent circuit with current sources replacing the conductance G_N and inductor L_J

$$G_N(V) = \begin{cases} 0, & |V| \leq 2\Delta_0/e \\ 1/R_N, & |V| \geq 2\Delta_0/e \end{cases} \tag{10.2}$$

All occurrences of V refer to the voltage across the three elements representing the Josephson junction from the node of their intersection to ground. The superconducting gap energy at zero temperature, Δ_0, gives the voltage where the junction transitions from superconducting to a normal metal with a normal resistance R_N, see Eq. (10.2). I_c is the critical current of the Josephson junction. This is the maximum current through the Josephson junction and sets the scale for the supercurrent, $I_c \sin \varphi$ in Eq. (10.2). For finite temperatures, Gross et al. [5] gives the temperature dependent conductance in the RCSJ based on the density of states of quasiparticles in the Josephson junction [5].

The VCCS $G_J(\varphi)$ varies sinusoidally with the junction phase, φ, which is a function of the voltage across the junction and given by the Josephson equation

$$V = \frac{\Phi_0}{2\pi}\frac{d\varphi}{dt} \tag{10.3}$$

$$\Phi_0 = \frac{h}{2e} \equiv \text{Magnetic flux quantum}$$

If we substitute Eq. (10.3) into Eq. (10.1), we arrive at a second order linear differential equation in the phase variable, φ

$$I_d(t) = \frac{\Phi_0}{2\pi}C_J\frac{d^2\varphi}{dt^2} + \frac{\Phi_0}{2\pi}\frac{d\varphi}{dt}G_N\left(\frac{\Phi_0}{2\pi}\frac{d\varphi}{dt}\right) + I_c \sin \varphi \tag{10.4}$$

This equation is analogous to a driven pendulum, where the capacitance and conductance are proportional to the mass and damping parameter for the pendulum, respectively [14]. For practical, classical simulations of Josephson junctions, the two VCCS model shunted by the junction capacitance is sufficient to produce hysteresis in the current-voltage (IV) characteristic curve. Numerical simulation of the circuit in Fig. 10.2 is well suited for SPICE [11] circuit solvers or coupled to geometries in multiphysics codes such as COMSOL.[1]

The RCSJ model is an intuitive model for the behavior of a Josephson junction with an applied dc or ac drive current, though it is not as suitable for superconducting qubit design and simulation. Circuit Quantum Electrodynamics provides a framework analyzing such systems with the language of atomic optics or cavity quantum electrodynamics. We will examine the key features of circuit QED and its utility in the design and simulation of superconducting qubits.

10.2.3   Circuit Quantum Electrodynamics

Circuit quantum electrodynamics (QED) combines microwave engineering, circuit analysis, and quantum optics. Fabry-Perot cavities from optics are replaced by resonant microwave cavities or lumped element microwave resonators in circuit QED. The procedure for obtaining the quantized Hamiltonian and subsequent dynamics of the system follows first from a classical treatment, then quantization of the classical variables as operators and relating those operators to bosonic single-mode raising and lowering operators $\left\{ \hat{a}_i^{(\dagger)} \right\}$.

10.2.3.1   Quantizing the LC Oscillator

We return to the LC oscillator circuit in Fig. 10.1 and write the Lagrangian for the circuit in terms of the flux variable ϕ which is treated as the generalized coordinate for the system [4].

$$\mathcal{L}\left(\phi, \dot{\phi}\right) = \frac{1}{2}C\dot{\phi}^2 - \frac{1}{2L}\phi^2 \tag{10.5}$$

We treat the charge q on the capacitor as the conjugate momentum and perform a Legendre transformation to obtain the Hamiltonian as a function of both q and ϕ.

$$q = \frac{\partial \mathcal{L}}{\partial \dot{\phi}} = C\dot{\phi} \Longrightarrow \dot{\phi}^2 = q^2/C^2$$

$$\mathcal{H}\left(q, \phi\right) = \dot{\phi}q - \mathcal{L} = \frac{1}{2}C\dot{\phi}^2 + \frac{1}{2L}\phi^2$$

$$\mathcal{H} = \frac{1}{2C}q^2 + \frac{1}{2L}\phi^2 \tag{10.6}$$

---

[1]COMSOL Multiphysics, www.comsol.com.

Following the example in Chapter 3 of [14], the charge and flux variables are quantized by converting them to operators with the commutation relation $\left[\hat{\phi}, \hat{q}\right] = i\hbar$. If we take the resonance frequency of the LC circuit to be $\omega = (LC)^{-1/2}$ and replace $1/L$ in the potential term of the Hamiltonian, we arrive at the familiar form for a harmonic oscillator with mass C.

$$\mathcal{H} \to \hat{H} = \frac{\hat{q}^2}{2C} + \frac{1}{2}C\omega^2\hat{\phi}^2 \tag{10.7}$$

We define raising and lowering operators for this quantum harmonic oscillator in analogy to those used in the one-dimensional model and write the second quantized form of the Hamiltonian.

$$\hat{q} = -i\sqrt{\frac{\hbar\omega C}{2}}\left(\hat{a} - \hat{a}^\dagger\right), \quad \hat{\phi} = \sqrt{\frac{\hbar}{2\omega C}}\left(\hat{a} + \hat{a}^\dagger\right) \tag{10.8}$$

$$\hat{H} = \hbar\omega\left(\hat{a}^\dagger\hat{a} + 1/2\right) \tag{10.9}$$

10.2.3.2 Black Box Circuit Quantization

In the previous section, we covered a procedure for quantizing an LC oscillator circuit which leads to an approximate generalization for any device given its frequency dependent impedance function. This approach connects full wave electromagnetic simulations of microwave circuits to their quantum mechanical analogs in circuit QED. Given a single port S-parameter as a function of frequency, one can obtain the impedance at the port by the transformation

$$Z = (\mathbb{1} + S)(\mathbb{1} - S)^{-1} \tag{10.10}$$

$\mathbb{1} \equiv$ identity matrix with same dimensions as S

Following the *Black box quantization* methods outlined in [12, 15], the impedance function, $Z(\omega)$ can be expressed as a pole-residue expansion in the complex frequency $s = j\omega$, where $j = -\sqrt{-1}$, following the electrical engineering convention.

$$Z(s) = \sum_{k=1}^{M} \frac{r_k}{s - s_k} + d + es \tag{10.11}$$

$\{r_k = a_k + jb_k\} \equiv$ residues, $\quad \{s_k = \xi_k + j\omega_k\} \equiv$ poles

The above rational function can be obtained by a least squares fit of the original impedance using the Vector Fit software outlined in [6] and available at The Vector Fitting Web Site – SINTEF, https://www.sintef.no/projectweb/vectfit/. If we take the case where $d = 0$ and the pole at $s \to \infty$ vanishes or $e = 0$ and perform the following partial fraction expansion and approximation for the k-th term in the series and we find the k-th term is the impedance for a parallel RLC oscillator circuit.

$$Z_k(s) = \frac{r_k}{s - s_k} = \frac{r_k}{s - s_k} + \frac{r_k^*}{s - s_k^*} \simeq \frac{2a_k s}{s^2 - 2\xi_k s + \omega_k^2}$$

$$\implies Z_k(s) = \frac{\frac{\omega_k r_k}{Q_k} s}{s^2 + \frac{\omega_k}{Q_k} s + \omega_k^2} \quad (10.12)$$

$$\omega_k = (L_k C_k)^{-1/2}, \quad Q_k = \omega_k R_k C_k = -\omega_k/2\xi_k, \quad R_k = -a_k/\xi_k$$

The total impedance, $Z(s)$ is a series combination of RLC oscillators and if we take the dissipationless limit by ignoring the resistances, we can treat $Z(s)$ as a series combination of LC circuits and apply the same analysis from Sect. 10.2.3.1 to each subcircuit. If we shunt the resulting circuit, with a single Josephson junction, we can obtain a simple model for the Hamiltonian of a qubit coupled to a superconducting resonator with M-modes. For a full derivation of the non-linear components of the Hamiltonian $\hat{H}_{nl}$, see [12]; we reproduce the salient features here.

$$\hat{H} = \hat{H}_0 + \hat{H}_{nl} \quad (10.13)$$

$$\hat{H}_0 = \sum_i \hbar \omega_i \hat{a}_i^\dagger \hat{a}_i, \quad \hat{H}_{nl} = E_J \left(1 - \cos \hat{\varphi} - \frac{\hat{\varphi}^2}{2} \right) \quad (10.14)$$

$$\hat{H}_{nl} \approx -\frac{1}{2} \sum_i \alpha_i \hat{a}_i^{\dagger 2} \hat{a}_i^2 - \sum_{i \neq j} \chi_{ij} \hat{a}_i^\dagger \hat{a}_i \hat{a}_j^\dagger \hat{a}_j \quad (10.15)$$

$$\hat{\varphi} = \frac{2\pi}{\Phi_0} \sum_i \hat{\phi}_i = \frac{2\pi}{\Phi_0} \sum_i \sqrt{\frac{\hbar}{2\omega_i C_i}} \left(\hat{a}_i + \hat{a}_i^\dagger \right) \quad (10.16)$$

The Hamiltonian above is referred to as the dispersive Hamiltonian for a weakly anharmonic qubit coupled to a series of harmonic modes. In the non-linear term, $\hat{H}_{nl}$, the first contribution describes the anharmonicities of those modes and the

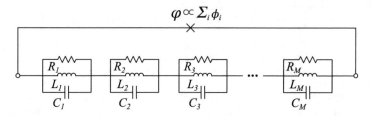

Fig. 10.3 Series combination of RLC circuits shunted by a single Josephson junction representing the black box circuit from Eq. (10.12) and similar in design to the circuit in [3]

qubit mode or self-Kerr terms and the second term gives the cross-Kerr terms [12]. Both $\{\alpha_i\}$ and $\{\chi_{ij}\}$ are experimentally observable, tying this model for qubit-circuit interactions to physical devices (Fig. 10.3).

10.3 Summary

The models used to describe the operation of superconducting qubits follow intuitive modifications to the familiar damped and driven oscillator systems from classical and quantum mechanics. These models arise from careful application of circuit QED to incorporate the quantum effects of macroscopic structures in microwave circuits. Although the dispersive Hamiltonian describes many superconducting qubit systems in quantum information experiments, this article did not apply the model to the problem of single photon counting. For more resources related to circuit QED and single photon counting, please refer to [1, 4, 7, 9, 13, 14].

Acknowledgements This work was performed under the auspices of the U.S. Department of Energy by Lawrence Livermore National Laboratory under Contract DE-AC52-07NA27344 and funded by the Laboratory Directed Research and Development programs at LLNL project numbers 15-ERD-051, 16-SI-004.

References

1. L. Bishop, Circuit quantum electrodynamics, Ph.D. thesis, Yale University, 2010
2. A. Blais, R.-S. Huang, A. Wallraff, S.M. Girvin, R.J. Schoelkopf, Cavity quantum electrodynamics for superconducting electrical circuits: an architecture for quantum computation. Phys. Rev. A **69**, 062320 (2004)
3. J. Bourassa, F. Beaudoin, J.M. Gambetta, A. Blais, Josephson-junction-embedded transmission-line resonators: from Kerr medium to in-line transmon. Phys. Rev. A **86**, 013814 (2012)
4. A.A. Clerk, M.H. Devoret, S.M. Girvin, F. Marquardt, R.J. Schoelkopf, Introduction to quantum noise, measurement, and amplification. Rev. Mod. Phys. **82**, 1155–1208 (2010)
5. R. Gross, A. Marx, F. Deppe, *Applied Superconductivity: Josephson Effect and Superconducting Electronics* (Walter de Gruyter, Berlin, 2016)
6. B. Gustavsen, A. Semlyen, Rational approximation of frequency domain responses by vector fitting. IEEE Trans. Power Delivery **14**(3), 1052–1061 (1999)
7. E. Holland, Cavity state reservoir engineering in circuit quantum electrodynamics, Ph.D. thesis, Yale University, 2015
8. E.T. Jaynes, F.W. Cummings, Comparison of quantum and semiclassical radiation theories with application to the beam maser. Proc. IEEE **51**(1), 89–109 (1963)
9. B. Johnson, Controlling photons in superconducting electrical circuits, Ph.D. thesis, Yale University, 2011
10. J. Koch, T.M. Yu, J. Gambetta, A.A. Houck, D.I. Schuster, J. Majer, A. Blais, M.H. Devoret, S.M. Girvin, R.J. Schoelkopf, Charge-insensitive qubit design derived from the Cooper pair box. Phys. Rev. A **76**, 042319 (2007)

11. L.W. Nagel, D. Pederson, SPICE (Simulation Program with Integrated Circuit Emphasis), Technical Report UCB/ERL M382, EECS Department, University of California, Berkeley, April 1973
12. S.E. Nigg, H. Paik, B. Vlastakis, G. Kirchmair, S. Shankar, L. Frunzio, M.H. Devoret, R.J. Schoelkopf, S.M. Girvin, Black-box superconducting circuit quantization. Phys. Rev. Lett. **108**, 240502 (2012)
13. M. Reed, Entanglement and quantum error correction with superconducting qubits, Ph.D. thesis, Yale University, 2013
14. D. Schuster, Circuit quantum electrodynamics, Ph.D. thesis, Yale University, 2007
15. F. Solgun, D.W. Abraham, D.P. DiVincenzo, Blackbox quantization of superconducting circuits using exact impedance synthesis. Phys. Rev. B **90**, 134504 (2014)

Chapter 11
Detecting Axion Dark Matter with Superconducting Qubits

Akash Dixit, Aaron Chou, and David Schuster

Abstract Axion dark matter haloscopes aim to detect dark matter axions converting to single photons in resonant cavities bathed in a uniform magnetic field. A qubit (two level system) operating as a single microwave photon detector is a viable readout system for such detectors and may offer advantages over the quantum limited amplifiers currently used. When weakly coupled to the detection cavity, the qubit transition frequency is shifted by an amount proportional to the cavity photon number. Through spectroscopy of the qubit, the frequency shift is measured and the cavity occupation number is extracted. At low enough temperatures, this would allow sensitivities exceeding that of the standard quantum limit.

Keywords Axion dark matter · Quantum non demolition · Single photon · Superconducting qubit

11.1 Single Photon Readout

In a haloscope search for axions, the signal signature is a population of a cavity mode with resonant frequency equal to axion mass and electric field with spatial overlap with the external magnetic field. Readout of this signal is enabled by quantum limited amplifiers which measure the power stored in the microwave cavity. The quantum limit manifests itself as zero point fluctuations of the resonator modes setting an irreducible noise floor. Searches for higher mass axions ($\geq 10\,\text{GHz}$) require smaller wavelength cavities whose volume shrinks as λ^3 resulting in a lower

A. Dixit (✉)
Department of Physics, University of Chicago, Chicago, IL, USA
e-mail: avdixit@uchicago.edu

A. Chou
Fermilab, Batavia, IL, USA

D. Schuster
University of Chicago, Chicago, IL, USA

© Springer International Publishing AG, part of Springer Nature 2018   97
G. Carosi et al. (eds.), *Microwave Cavities and Detectors for Axion Research*,
Springer Proceedings in Physics 211, https://doi.org/10.1007/978-3-319-92726-8_11

signal rate. Additionally, the noise rate associated with zero point fluctuations scales linearly with frequency (ω). The signal to noise ratio drops rapidly such that the scan rate becomes untenable. To increase the signal the target volume can be increased possibly by power combining multiple smaller cavities or increasing the external magnetic field. Each of these presents significant challenges in complexity and cost. The other possibility is to implement a readout scheme that is insensitive to the noise floor by measuring only a single quadrature of the cavity field. Rather than measuring power stored in the cavity mode, counting the photon population provides the information about the axion mass without incurring the uncertainty added by a linear amplifier operating at the quantum limit [1, 2]. In accordance with the Heisenberg uncertainty principle, the back action from this noiseless amplitude measurement results in the randomization of the photon phase. However, the phase provides no information about the mass of the axion and increased fluctuations in this quadrature do not degrade the measurement.

11.2   Cavity Quantum Electrodynamics

A single photon measurement can be made using the interaction between the resonator and a qubit (two level system). The physics of this interaction is governed by the Jaynes-Cummings Hamiltonian. In second order perturbation theory with qubit-cavity coupling (g) much smaller than both the cavity (ω_c), qubit (ω_q) transitions the Hamiltonian is given by:

$$\mathscr{H} = \omega_c a^\dagger a + \omega_q \sigma_z + \frac{g^2}{\Delta} a^\dagger a \sigma_z \tag{11.1}$$

Δ is the difference in the cavity and qubit frequencies. The interaction term contains only number operators and thus commutes with the unperturbed Hamiltonian. This allows us to perform quantum non demolition measurements of the cavity photon number (or qubit population). This effect manifests itself as a frequency shift of the qubit (cavity) transition as a function of cavity photon (qubit) population.

11.3   Microwave Cavity

A standard cavity, compatible with e solenoid magnet, at $\sim 10\,\text{GHz}$ would employ the TM010-like mode of a right cylindrical cavity ($r = 12\,\text{mm}$). For integration with a superconducting qubit, the geometry must be modified such that the sensor is contained in a region free of magnetic flux. This is achieved by aperture coupling another long right cylindrical cavity with slightly larger radius ($r = 12.1\,\text{mm}$) to the standard cavity. The lowest order mode ($9.52\,\text{GHz}$) of this composite system can be approximated as the TM010 of the larger radius cavity which has minimal coupling to axion induced field but maximal coupling to the qubit making it ideal to **readout** the qubit state. The next mode (9.65 GHz) has nonzero coupling to both the axion field and qubit and will serve as the **detection** mode.

11.4 Transmon Qubit Fabrication

Qubit frequency and coupling to cavity modes can be engineered through simulation with HFSS and the blackbox calculator [3]. Qubit fabrication is a multistep process done either on a Silicon or Sapphire substrate. Optical lithography of a layer of Niobium produces the pads that set the charging energy, E_C, and the coupling to the cavity. Electron beam lithography is used to pattern channels for the Josephson junction. Finally, two layers of aluminum are deposited in an angled evaporator with an oxidation step in between to create the tunnel junction (AlOx) approximately 1 nm in thickness (see Fig. 11.1). The junction area and oxide thickness set the Josephson energy, E_J [4, 5]. Designing the cooper pair box to have $\frac{E_J}{E_C} \sim 100$ provides stability from charge noise and is termed a transmon qubit [6].

11.5 Experimental Setup

The qubit is mounted into the cavity by placing the chip into a slot through the cap. The chip contacts the copper of the cavity on one side and is held in the slot with GE varnish applied to the other side of the chip (see Fig. 11.2). The cavity +

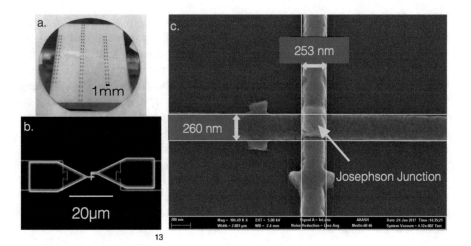

Fig. 11.1 (a) Wafer of optically patterned capacitor pads (b) Electron beam lithography (c) SEM image of double angle evaporated aluminum forming Josephson junction

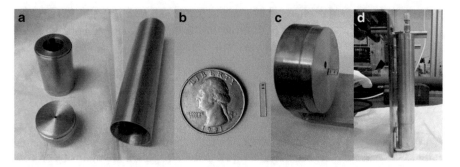

Fig. 11.2 (**a**) Microwave cavity (**b**) Qubit on sapphire substrate (**c**) Qubit chip loaded into cavity cap (**d**) Qubit-cavity system mounted onto copper plate

qubit system is placed in a μ-metal shield and bolted to the dilution refrigerator operating at 20 mK. The input to the cavity contains multiple attenuators including an eccosorb filter to cutoff high frequency radiation. Finally a direction coupler connects to the single antenna into the cavity which interacts with both the readout and detection mode as well as the qubit. The reflection from the antenna passes through the directional coupler into two circulators and amplified by a Low Noise factory amplifier at 4 K. Measurements at room temperature are done with network analyzer (CW) or a homodyne setup (Pulsed) including RF sources and an arbitrary wave form generator.

11.6   Results

The interaction between qubit and cavity modes results in nonlinear behavior or the cavity modes which can be probed by varying the probe power and observing the resonator frequency shift (Fig. 11.3). Driving the qubit with a pulse and allowing the excited state population decay provides a measure of the qubit lifetime (T1 1 μs) (Fig. 11.4). Varying the pulse length results in the qubit population Rabi flopping (see Fig. 11.5). The current T1 seem to be limited by the strong coupling to the cavity modes resulting is larger loss rates. This hypothesis is currently being tested by implementing the qubit system in a cavity configuration with lower mode density (less modes to couple to) and smaller dipole arm (less coupling to modes). Additionally the excited state of the qubit has a non zero residual population which would result in a false positive signal in our scheme. Attempts to mitigate this excited state population include a clamping mechanism to well thermalize the qubit and applying magnetic flux to the capacitor pads to trap potentially excited quasi particles in the Niobium.

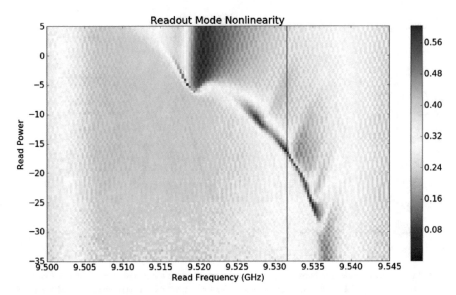

Fig. 11.3 This scan shows the cavity spectrum as a function of probe power (*y*-axis in dB). The color scale indicates the depth of the cavity resonance as measured at the digitizer. The resonator coupled to the qubit exhibits a characteristic Jaynes-Cummings nonlinearity. Increasing the probe power produces a duffing type response until the cavity frequency snaps at high photon number

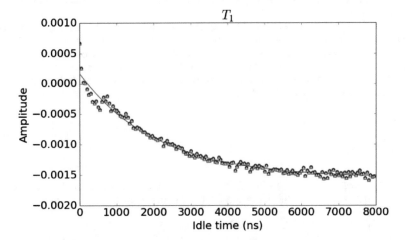

Fig. 11.4 Qubit population is measured at various delay times to extract the characteristic T1 decay time

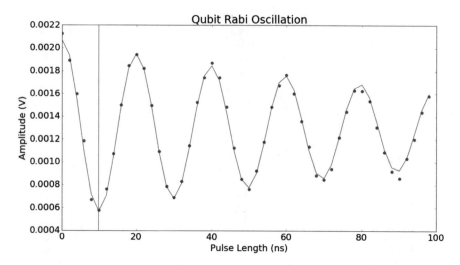

Fig. 11.5 Rabi oscillations between ground and excited qubit states are stimulated by driving at the qubit frequency. The qubit population is observed as it traverses the Bloch sphere. The vertical line at 10 ns indicates where the qubit has reached the excited state

Acknowledgements This work made use of the Pritzker Nanofabrication Facility of the Institute for Molecular Engineering at the University of Chicago, which receives support from SHyNE, a node of the National Science Foundations National Nanotechnology Coordinated Infrastructure (NSF NNCI-1542205).

This manuscript has been authored by Fermi Research Alliance, LLC under Contract No. DE-AC02-07CH11359 with the U.S. Department of Energy, Office of Science, Office of High Energy Physics. The United States Government retains and the publisher, by accepting the article for publication, acknowledges that the United States Government retains a non-exclusive, paid-up, irrevocable, world-wide license to publish or reproduce the published form of this manuscript, or allow others to do so, for United States Government purposes.

References

1. S.K. Lamoreaux, K.A. van Bibber, K.W. Lehnert, G. Carosi, Analysis of single-photon and linear amplifier detectors for microwave cavity dark matter axion searches. Phys. Rev. D **88**, 035020 (2013)
2. T.M. Shokair, J. Root, K.A. van Bibber, B. Brubaker, Y.V. Gurevich, S.B. Cahn, S.K. Lamoreaux, M.A. Anil, K.W. Lehnert, B.K. Mitchell, A. Reed, G. Carosi, Future directions in the microwave cavity search for dark matter axions. Int. J. Mod. Phys. A **29**, 1443004 (2014)
3. S.E. Nigg, H. Paik, B. Vlastakis, G. Kirchmair, S. Shankar, L. Frunzio, M.H. Devoret, R.J. Schoelkopf, S.M. Girvin, Black-box superconducting circuit quantization. Phys. Rev. Lett. **108**, 240502 (2012)
4. D.I. Schuster, Circuit quantum electrodynamics, Ph.D. thesis, Yale University, 2007

5. B.R. Johnson, Controlling photons in superconducting electrical circuits, Ph.D. thesis, Yale University, 2011
6. J. Koch, T.M. Yu, J. Gambetta, A.A. Houck, D.I. Schuster, J. Majer, A. Blais, M.H. Devoret, S.M. Girvin, R.J. Schoelkopf, Charge-insensitive qubit design derived from the cooper pair box. Phys. Rev. A **76**, 042319 (2007)

Chapter 12
Recent Technical Improvements to the HAYSTAC Experiment

L. Zhong, B. M. Brubaker, S. B. Cahn, and S. K. Lamoreaux

Abstract We report here several technical improvements to the HAYSTAC (Halo-scope at Yale Sensitive To Axion Cold dark matter) that have improved operational efficiency, sensitivity, and stability.

Keywords Axion · Cold dark matter · HAYSTAC · Haloscope · Josephson Parametric Amplifier (JPA) · Attocube motor tuning · Hot rod · Thermal link · Kevlar · Quality factor

12.1 Overview

The HAYSTAC (Haloscope at Yale Sensitive To Axion Cold dark matter) was commissioned in January 2016 and has been operational for well over a year. The experiment employs a high-Q tunable microwave cavity that is immersed in a strong magnetic field (9 T). Putative galactic halo axions convert to radiofrequency (RF) photons in the strong magnetic field, and the cavity serves as an (imperfect) impedance matching network that couples the near infinite impedance signal source to a coaxial cable (this can be understood as an extension of the Purcell effect, as originally conceived [1]), which in turn delivers the RF power to a Josephson Parametric Amplifier (JPA). Because the axion mass, hence RF frequency, is not known, the cavity and amplifier need to be tunable. The experiment thus requires a slow search over frequency for an excess RF noise signal due to axion conversion that would appear as an addition to expected quantum fluctuation noise (along with minimal thermal noise).

The first data run was completed in August 2016, with analysis results reported in [2] and a detailed description of the apparatus published in [3]. In the course of this data run, several problems were identified. First, the Kevlar pulley system that was

L. Zhong · B. M. Brubaker · S. B. Cahn · S. K. Lamoreaux (✉)
Yale University, Physics Department, New Haven, CT, USA
e-mail: l.zhong@yale.edu; steve.lamoreaux@yale.edu

© Springer International Publishing AG, part of Springer Nature 2018
G. Carosi et al. (eds.), *Microwave Cavities and Detectors for Axion Research*,
Springer Proceedings in Physics 211, https://doi.org/10.1007/978-3-319-92726-8_12

used to vary the cavity resonance frequency by rotating the tuning rod internal to the cavity had considerable hysteresis and would drift over about 20 min after taking a small frequency step (typically 100 kHz, corresponding to 0.003° rotation), due mainly to stiction in the tuning rod bearings. Second, the tuning rod was supported solely by thin alumina tubes that did not provide a sufficient thermal link to the tuning rod which ended up stuck at a temperature around 600 mK, far above the base temperature of 125 mK. Finally, the use of thick Cu elements in the construction of the cavity support framework led to major damage of the experiment from the eddy current forces resulting from a superconducting magnet quench during a power outage in March 2016. In this note we will describe improvements to the experiment that address these problems.

12.2   Attocube Motor Tuning

Because of the time-dependent drift after stepping the frequency using the Kevlar line pulley tuning system, we only made large frequency steps with that system, and used the insertion of a thin dielectric rod to perform the required fine stepping. Unfortunately, the range of tuning with the dielectric rod depends on frequency, and in some regions has no effect at all. Motion of the dielectric rod generated significantly more heat than the pulley system tuning.

To eliminate the stiction and hysteresis problems, we replaced the Kelvar line pulley tuning system with an Attocube ANR240 piezoelectric precision rotator (http://www.attocube.com/attomotion/premium-line/anr240/). The rotator is supported by a bracket attached to the experiment frame, about 12″ above the cavity. The rotary motion is transmitted by 6″ long 0.25″ brass rods, connected by a corrugated stainless flexible shaft coupler.

The rotator requires a sawtooth-voltage electrical signal (with amplitude 45 V) and draws a high current (around 1.5 A) when actuated. We therefore spatially separate the two wires required by the rotator from the delicate electronics wiring (flux bias current, HEMT amplifier controls, thermometry, heater and microwave switch) and provide a separate vacuum feedthrough. The resistance of the rotator wiring must be kept low, so we used 28 AWG Cu from room temperature to the 4 K stage, and NbTi superconducting wire below the 4 K stage. Because the superconducting wire has too much series resistance at room temperature, the rotator cannot be tested as-wired before cooling down the system. Room temperature tests of the rotator alone are possible by use of a temporary all-Cu low resistance wire pair.

The Attocube rotator has sufficient torque to move the cavity tuning rod even in the presence of the 9 T field. Empirically the mechanical stiction depends on the direction of rotation. At 9 T, we find that we are only able to tune in one direction at certain angles where the stiction is large. However, unidirectional tuning is sufficient for an axion search, and we can tune freely in both directions by reducing the magnetic field to 6 T.

The tuning system heat load which was mainly due to large dielectric rod motions with the previous system has been improved with the current system. It should be noted that the Kevlar line pulley system generates less heat than the Attocube system, however the latter has provided seamless operation with an acceptable heat load and no significant drift. It is likely that even if the bearing stiction problem is solved, a drift issue would remain with the pulley system, albeit at a much reduced and possibly acceptable level.

12.3 The Hot Rod Problem and Solution

During the commissioning phase, it was noted that the noise power at the cavity resonant frequency was always significantly higher than the noise power observed off-resonance. The off-resonance noise was consistent with the expected system noise temperature. Various tests, including raising the system temperature to the point where the cavity and off-resonance noise levels were almost equal, provided firm evidence that the excess noise was due to the tuning rod (or other system element) being at a high temperature. These tests alleviated concerns that the excess noise was due to a spurious interaction between the cavity and JPA which might have been difficult to eliminate. Such a large-volume and uniquely configured cavity had never before been coupled to a JPA.

We had long known that the thermal link to the cavity tuning rod was at best weak, being several inches of 0.250″ OD, 0.125″ ID polycrystalline alumina tubes, at least an inch long, on either end of the cavity cylinder axis. The only contact between the tubes and the support frame (which serves as the thermal link to the dilution refrigerator mixing chamber) were precision ball bearings that were used to ensure as free a rotation of the rod as practical. The contact area between the balls and the races is vanishingly small by design, further complicating the thermal link problem. We had recognized this problem earlier, and attempted to provide a link by gluing (using thermal epoxy) a short brass rods (0.25″) into the external ends of the alumina tubes, and then connecting those brass rods to the support frame with flexible Cu braids.

Tests at U.C. Berkeley suggested that a 0.125″ Cu rod could be inserted far into alumina tubes (to a depth where they are within the cavity), without undue loss of cavity Q (K. van Bibber and M. Simanovskaia, 2017, U.C. Berkeley, Private Communication). Such rods were incorporated, inserted as far as possible into the alumina tubes, and glued in place using conductive silver epoxy (Epo-Tek H20E). The rods are long enough so a 1/2″ nub extends beyond the tube, and Cu braids were soldered onto these nubs before gluing. The upper nub serves as the connection point to the piezoelectric rotator (discussed above).

According to the latest measurements, the Cu rods reduce the total system noise photon number at the cavity resonant frequency from around 3 to 2.3 on average, corresponding to a reduction in tuning rod temperature from 600 to 250 mK. Unfortunately, the cavity Q has been reduced by about 40%.

Incorporation of the Cu rod thermal links reduced the time to cool the system from when the mixture is first condensed to the base temperature from over 6 h to under 1 h. This was because the alumina tubes' weak thermal link became weaker with reduction in temperature, becoming a bottleneck in the cooling process while maintaining a substantial heat load. The alumina tubes became effective thermal insulators when the end affixed to the tuning rod reached 600 mK (with the other end at the 125 mK frame temperature). After this quasi-equilibrium was established, we saw no discernible decrease in thermal noise level over months of operation, implying a time of perhaps years for the tuning rod to significantly cool beyond this point. We have yet to identify the remaining source of excess thermal noise (250 mK compared to a system temperature of 125 mK) which is likely a further issue with the tuning rod thermal link.

An unfortunate consequence of the thermal links is that noise is coupled into the cavity. The links act as antennas that couple signals directly into the cavity, as might be expected because the mechanisms that result in reduction in Q likely have a component due to the internal field leaking, and radiating, to free space. Although one might expect the radiofrequency (RF) noise within the cryostat to be extremely low, there apparently is enough noise (mostly from coherent sources, so technically it is not noise but spurious or systematic signals) to be problematic. There are many possible noise sources, including coupling between the room temperature part of the cryostat and the cavity region due to signals running on thermometer wires, the outsides of coaxial cables, or the cryostat support structure and piping. RF noise can also couple into the cryostat through the myriad thermometer, heater, and control wires. Although there are filters in place that offer more than 100 dB of attenuation, we must recall that the system is sensitive at the single photon or Yoctowatt level. The current operating frequency of 5.6–5.7 GHz is the so-called 5 GHz WiFi band, and our impression is that this is the source of most of the noise.

The RF noise problem can be solved by enclosing the thermal links in a cylindrical shield that attaches to the cavity and surrounds the rods, with the flexible braid attached to the inner surface of the shield, and an endcap on the cylinder.

12.4   Cu Plated Stainless Thermal Links and Shields

Our original design incorporated many massive OFHC Cu components, and as mentioned in the introduction, the forces that were generated during a magnet quench led to significant damage to the experiment. We have noticed that the cavity, fabricated from stainless steel and Cu plated, apparently has a very high thermal conductivity. Thermometers are mounted on the cavity top (where the thermal link to the frame is located) and bottom, and practically there is almost no time delay (less than 30 s) between a temperature change at the top, and subsequent change at the bottom.

The damaged still-temperature thermal shield was replaced with a Cu plated (to 0.002″) stainless steel shield (the manufacturing and plating of the shield was arranged by K. van Bibber of U.C. Berkeley). This new shield has been sufficient with no obvious excess heat load at the mixing chamber level. We have yet to see its response to a magnet quench, however we plan to replace all massive Cu parts with Cu plated stainless steel in the next upgrade to the experiment.

12.5 Conclusion

Incorporation of the Attocube rotator has solved the hysteresis and drift problem of previous Kevlar pulley system, and reduced the complexity of operation. The addition of the Cu rods has provided better cooling efficiency of the cavity tuning rod at the cost of 40% reduction of cavity Q: the net result is that operation of the system is more reliable and the scan rate is unchanged, however the rods allow RF noise to be coupled into the cavity so there are additional peaks that require extra and time-consuming attention; this problem does have a solution. In the near future, we plan to upgrade the experiment, which will include moving it to a new BlueFors dilution refrigerator, the replacement of massive Cu components in the high magnetic field regions with Cu plated stainless steel, and the incorporation of a squeezed state receiver system as describe in [4].

Acknowledgements This work was supported by the National Science Foundation, under grants PHY-1362305 and PHY-1607417, by the Heising-Simons Foundation under grants 2014-181, 2014-182, and 2014-183, and by Yale University. HAYSTAC is a collaboration between U.C. Berkeley, Colorado University/JILA, and Yale.

References

1. E.M. Purcell, Phys. Rev. **69**, 681 (1946)
2. B.M. Brubaker et al., Phys. Rev. Lett. **118**, 061302 (2017)
3. S. Al Kenany et al., Nucl. Instrum. Methods **854**, 11 (2017)
4. H. Zheng et al., arXiv:1607.02529

Chapter 13
Multiple-Cavity Detector for Axion Search

Sung Woo Youn

Abstract Searching higher frequency regions for axion dark matter using microwave cavity detectors requires smaller size cavities as the resonant frequencies scale inversely with cavity radius. One of the intuitive ways to make an efficient use of a given magnet volume is to bundle an array of cavities together and combine their individual outputs ensuring phase-matching of the coherent axion signal. In this article, an extensive study of realistic design for the phase-matching mechanism of multiple-cavity systems is performed and its experimental feasibility is demonstrated using a double-cavity system.

Keywords Axion · Dark-matter · Multi-cavity · Sensitivity · Combiners · Frequency matching · Tuning · Demonstration · Alumina rod · 5 GHz

13.1 Introduction

The axion, motivated by the Peccei-Quinn mechanism to solve to the CP problem in quantum chromodynamics of particle physics [1], is an attractive cold dark matter candidate [2]. The current detection method, suggested by Sikivie, utilizes microwave resonant cavities placed in a strong magnet where axions are converted to radio-frequency (RF) photons [3]. Cavity-based axion search experiments typically employ a single resonant cavity fitting into a given magnet bore. However, exploring higher frequency regions requires smaller cavity sizes as the frequency of the resonant mode of our main interest, TM_{010}, is inversely proportional to cavity radius R, i.e., $f_{TM_{010}} \sim R^{-1}$. An intuitive way to increase the detection volume for the given magnet, and thereby to increase the sensitivity, is to bundle multiple cavities together and combine the individual outputs coherently, which is referred to as "phase-matching". In this article, we present an extensive study, where

S. W. Youn (✉)
Center for Axion and Precision Physics Research, Institute for Basic Science, Daejeon, South Korea
e-mail: swyoun@ibs.re.kr

© Springer International Publishing AG, part of Springer Nature 2018
G. Carosi et al. (eds.), *Microwave Cavities and Detectors for Axion Research*,
Springer Proceedings in Physics 211, https://doi.org/10.1007/978-3-319-92726-8_13

a conceptual design of phase-matching mechanism is driven, and demonstrate its experimental feasibility using a double-cavity detector. More detailed description is found at Ref. [4].

13.2 Configurations

There are three possible configurations in designing the experimental setup for a multiple(N)-cavity system as summarized in Table 13.1. Configuration 1 comprises N single-cavity experiments, consisting of a complete receiver chain per cavity, where the signals are statistically combined after all, eventually resulting in a $\sqrt{N}$ improvement in sensitivity. The other two configurations introduce a power combiner at an early stage of the receiver chain to build an N-cavity experiment— one with the first stage amplification taking place before the signal combination; and the other with the signal combination preceding the first stage amplification. Configuration 2 is characterized by N amplifiers and a combiner, while configuration 3 is characterized by a single amplifier and a combiner. Assuming the axion signal from individual cavities is correlated while the noise from system components is uncorrelated, configuration 2 gains an additional $\sqrt{N}$ improvement yielding the highest sensitivity. On the other hand, configuration 3 provides the simplest design with a compatible sensitivity with that of configuration 2. As simpler design is significantly beneficial especially for large cavity multiplicities, configuration 3 is chosen as the final design.

13.3 Phase-Matching

13.3.1 Frequency-Matching

Due to the large de Broglie wavelength of the coherent axion field and relatively facile achievement of constructive interference in signal combination, the phase-matching of a multiple-cavity system is in practice equivalent to frequency tuning of individual cavities to the same resonant frequency, which is referred to as frequency-matching. Unfortunately, an ideal frequency-matching of multiple cavities is not possible mainly because of the machining tolerance of cavity fabrication and non-zero step size of the turning system.[1] Instead, a more realistic approach is to permit frequency mismatch up to a certain level at which the combined power is still sufficiently enough that the resulting sensitivity is not significantly degraded. We refer to the certain level as the frequency matching tolerance (FMT).

---

[1] Typical values of machining tolerance and step size of piezoelectric rotators are 50 μm and 0.1 m°, respectively. They correspond to a frequency difference of $\sim$10 MHz and a frequency step of $\sim$0.5 kHz for the TM$_{010}$ mode of a 5 GHz resonant cavity.

Table 13.1 Possible configurations of the receiver chain for a multiple(N)-cavity system

Configuration	1	2	3
Schematic			
Characteristics	N complete chains	N amplifiers 1 combiner	1 amplifier 1 combiner
Sensitivity	$\sqrt{N} \cdot \mathrm{SNR_{sgl}}$	$N \cdot \mathrm{SNR_{sgl}}$	$N \cdot \mathrm{SNR_{sgl}}$
Pros	Accessibility to individual cavities	Highest sensitivity	Simplest design
Cons	Low sensitivity complex design	N amplifiers	$\mathrm{SNR_3} \lesssim \mathrm{SNR_2}$[a]

The cylinders, triangles, and D-shaped figures represent cavities, amplifiers, and combiners, respectively. $\mathrm{SNR_{sgl}}$ refers to the signal-to-noise ratio (SNR) of a single-cavity experiment. The gain of the amplifiers is assumed to be large enough

[a]Configuration 3, where the combiner is placed prior to the first amplifier, degrades the SNR due to imperfection of the combiner. For instance, a system with a combiner with a noise figure of 0.5 and a amplifier with a gain of 12 and a noise figure of 6 yields a SNR reduction of $\sim 10\%$

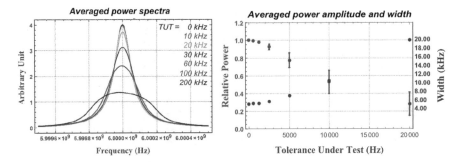

Fig. 13.1 Combined power spectra averaged over 1000 pseudo-experiments for several TUT values (left). The power amplitude from each cavity is normalized to the unity. Relative power amplitude in red and full width at half maximum in blue as a function of TUT (right). The error bars represent the statistical uncertainties. These distributions are fitted with the Lorentzian and fourth order polynomial functions, respectively

13.3.2 Frequency Matching Tolerance

To determine FMT for multiple-cavity systems, a pseudo-experiment study is performed using a quadruple-cavity detector searching for 5 GHz axion signal. The unloaded quality factors of all cavities are assumed to be the same as $Q_0 = 10^5$. Supposing $Q_a \gg Q_0$, the signal power spectrum follows the Lorentzian distribution with its mean of 5 GHz and half width of 50 kHz. Several values of frequency matching tolerance, tolerance under test (TUT), are considered, i.e., 0, 5, 10, 20, 30, 60, 100, and 200 kHz, where 0 kHz corresponds to the ideal frequency-matching. The cavities in the array are randomly tuned to the target frequency, 5 GHz, following a uniform distribution with its center at 5 GHz and half-width of the TUT under consideration. Assuming each cavity is critically coupled, the individual power spectra are linearly summed up. The combined power spectrum is fitted with the Lorentzian function to get the amplitude and full-width at half maximum. The procedure is repeated 1000 times over which the combined power spectra are averaged. Figure 13.1 shows the distributions of the averaged combined power spectra for different TUT values (left) and displays the normalized amplitudes and full widths at half maximum as a function of TUT (right). For the realistic approach, we put criteria that the relative amplitude of the combined power spectrum is greater than 0.95. From this we find that the FMT is 21 kHz for a system consisting of four identical cavities with $Q_0 = 10^5$ seeking for a 5 GHz axion signal. The FMT has a dependence on the cavity quality factor and target frequency, and can be generalized as $\mathrm{FMT}(Q_0, f) = 0.42\,\mathrm{GHz}/Q_0 \times f$ [GHz].

13.4 Tuning Mechanism

The basic principle of the tuning mechanism for a multiple-cavity system employs the same principle as for conventional single-cavity experiments. It relies on target frequency shift by rotating a single dielectric rod inside the cavity; frequency matching by finely tuning the individual cavity frequencies; and critical coupling by adjusting the depth of a single RF antenna into the cavity. Frequency-matching is assured by the minimum bandwidth of the reflection peak giving the maximum Q value. Critical coupling is characterized by the minimum reflection coefficient (Γ) in the scattering parameter and by the constant resistance circle passing through the center of the Smith chart [5]. Therefore, the tuning mechanism for a multiple-cavity system consists of three steps: (1) shifting the target frequency by simultaneously operating the rotational actuators; (2) achieving frequency matching by finely manipulating the individual rotational actuators; and (3) achieving critical coupling of the system by globally adjusting the antenna depth using the linear actuator.

13.5 Experimental Demonstration

The feasibility of the tuning mechanism for multiple-cavity systems is experimentally demonstrated using a double-cavity detector at room temperature. It is composed of two identical copper cavities with an inner diameter of 38.8 mm whose corresponding resonant frequency is 5.92 GHz and an unloaded quality factor of about 18,000. A single dielectric rod made of 95% aluminium oxide (Al_2O_3) with 4 mm diameter is introduced to each cavity and a piezoelectric rotator is installed under the cavity to rotate the rod for frequency tuning. With the tuning rod positioned at the center of the cavity, the resonant frequency decreased to 4.54 GHz and Q_0 degrades to about 5,000 due to energy loss by the rod. A pair of RF antennae, each of which is coupled to each cavity, sustained by an aluminium holder attached to a linear piezoelectric actuator above the cavities.

The double-cavity system is assembled by connecting the two antennae to a two-way power combiner which transmits signals to a network analyzer. Critical coupling of one cavity is made while the combiner input port, which the other cavity is connected to, is terminated with a 50 Ω impedance terminator, and vice versa. Two cavities are configured to be critically coupled at slightly different resonant frequencies. The initial values of the loaded quality factor Q_L and scattering parameter (S-parameter) S_{11} are measured. After the system is re-assembled with the both RF antennae being connected to the combiner, the initial system is represented by two reflection peaks with $S_{11} = -6$ dB in the S-parameter spectrum[2] and two small circles with half-unit radius on the Smith chart.

In order to match the frequency, one of the rotational actuators is finely manipulated until the combined reflection coefficient becomes minimized. Following that, the linear actuator is operated to adjust the antenna positions in a global manner to achieve critical coupling of the system until the reflection peak becomes the deepest and the large circle passes through the center of the Smith chart. The final

---

[2]Two cavities are on resonance at different frequencies at which the combiner does not see the reflected signal. Note that for the reflected signal the combiner acts as a splitter so that the final reflected signal at the common port is only one-fourth of the input signal.

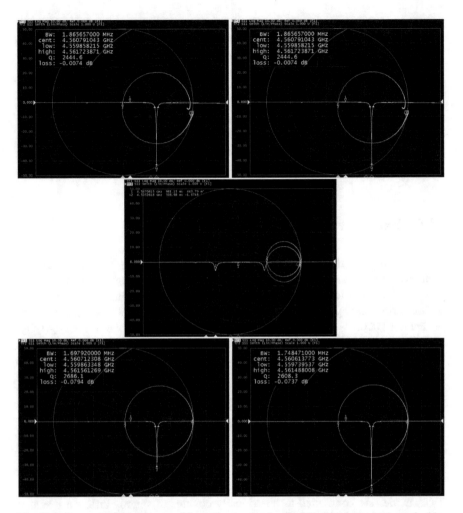

Fig. 13.2 Demonstration sequence of the tuning mechanism described in the text using a double-cavity system. Yellow solid lines are the scattering parameter, S_{11}, spectra in a logarithmic scale and cyan circles are their representations in the Smith chart. The first (second) cavity is critically coupled separately, and the initial S_{11} and Q_L read −46.4 (−43.2) dB and 2,530 (2,440) respectively (top left (right)). Two −6 dB peaks and two small circles are observed after the system is fully assembled (middle). When the frequency is matched, a single deep peak and a single large circle are formed (bottom left). Once critical coupling is accomplished in a global way, the peak becomes deeper and the circle passes through the center of the Smith chart. The final S_{11} and Q_L read −44.4 dB and 2,610 (bottom right)

Q_L and S_{11} values are measured. The consistency between initial and final values convinces that the turning mechanism with frequency-matching is successful. The demonstration sequence and characteristics of each step are shown in Fig. 13.2.

13.6 Conclusions

We performed an extensive study of multiple-cavity design as an effective way
to increase the sensitivity of axion search experiments in high frequency regions.
A receiver chain with signal combination preceding the first stage amplification
is beneficial because of the simplest design with minimal degradation of signal
power. The frequency-matching between individual cavities is a key component of
the phase-matching mechanism. For a realistic approach, the frequency matching
tolerance is introduced and numerically determined for a quadruple-cavity detector
for 5 GHz axion signal. An experimental demonstration of the tuning mechanism
(frequency matching and critical coupling), successfully conducted using a double-
cavity detector, verifies its certain application to axion search experiments.

Acknowledgements This work was supported by IBS-R017-D1-2017-a00/IBS-R017-Y1-2017-
a00.

References

1. R.D. Peccei, H.R. Quinn, Phys. Rev. Lett. **38**, 1440 (1977); R.D. Peccei, H.R. Quinn, Phys. Rev.
   D **16**, 1791 (1977)
2. J. Preskill, M.B. Wise, F. Wilczek, Phys. Lett. B **120**, 127 (1983)
3. P. Sikivie, Phys. Rev. Lett. **51**, 1415 (1983)
4. J. Jeong et al., Astropart. Phys. **97**, 33 (2018)
5. P.H. Smith, Electronics **12**, 29 (1939); P.H. Smith, Electronics, **17**, 130 (1944)

Chapter 14
The ORGAN Experiment

Ben T. McAllister and Michael E. Tobar

Abstract The **O**scillating **R**esonant **G**roup **A**xio**N** experiment (ORGAN), is a haloscope search for high mass axions hosted at the University of Western Australia node of the ARC Centre of Excellence for Engineered Quantum Systems (EQuS). The experiment has received 7 years of funding through EQuS, and will be a collaboration of the various EQuS nodes. We discuss the targeted parameter space of the search, search methodology, some novel resonator design and a scheme for power combining resonators.

Keywords Axion · Cold dark matter · ORGAN · Australia · Quantum · Pathfinder · TM020 · Dielectric resonators · Cross-correlation · Cavities

14.1 High Mass Haloscopes

The push towards high mass ($>$ 10 μeV) axion haloscope searches is the subject of much contemporary work [1–5]. Several recent theoretical developments point towards the probability of axions above the mass ranges that are typically searched in direct detection experiments. The recent SMASH model [6] predicts axions between 50 and 200 μeV, with a preferred mass of 100 μeV, whilst another recent paper suggests that anomalous signals in Josephson junctions can be explained as a result of axions entering the junction and transporting excess Cooper pairs [7]. The paper claims that axions with a mass of 110 μeV can explain these signals. Simulations of lattice QCD suggest axions with a mass greater than 14 μeV, but less than 10^4 μeV [8]. These results and others combine to motivate high frequency haloscope searches to directly probe this axion-photon coupling parameter space. Despite this, due to a host of difficulties associated with high mass haloscope

B. T. McAllister (✉) · M. E. Tobar
ARC Centre of Excellence for Engineered Quantum Systems, School of Physics,
University of Western Australia, Crawley, WA, Australia
e-mail: ben.mcallister@uwa.edu.au; michael.tobar@uwa.edu.au

© Springer International Publishing AG, part of Springer Nature 2018   119
G. Carosi et al. (eds.), *Microwave Cavities and Detectors for Axion Research*,
Springer Proceedings in Physics 211, https://doi.org/10.1007/978-3-319-92726-8_14

searches, no direct search above $25\,\mu eV$, with better sensitivity than that of CAST has ever been performed.

14.2   ORGAN

The ORGAN experiment is a haloscope search undertaken as a collaboration of several nodes of the ARC Centre of Excellence for Engineered Quantum Systems (EQuS). The experiment will search for axions in the broad range of 15–50 GHz, with an initial focus around 26.6 GHz to provide a direct test of the claimed result discussed above. The experiment recently underwent a stationary frequency pathfinding run, designed to test apparatus for future runs, and to develop some institutional experience with haloscope searches.

14.2.1   Pathfinding Run

The ORGAN Pathfinder experiment consisted of a small copper resonant cavity and a traditional HEMT-based amplifier (effective noise temperature $\sim$8 K) held at 4 K, embedded in a 7 T superconducting magnet held at 30 mK. The cavity had no tuning mechanism, and the TM_{020} mode frequency of the cavity was 26.531 GHz, with a loaded quality factor of $\sim$13,000. The run was a success in that it demonstrated the ability of the system to collect data for extended periods of time, and for the readout and data processing chains to operate as designed. The narrow 95% confidence exclusion limits on axions are shows in Fig. 14.1.

14.2.2   Future Experiments

The experiment has recently received 7 years of funding through EQuS. As such, we will perform a series of long-term data collection experiments, each targeting a different region of the promising, unprobed axion parameter space.

14.2.2.1   Stage I

The first stage, planned for the duration of 2018 will utilize two small cavities, embedded in a 14 T magnet at 30 mK, with traditional HEMT based amplification at 4 K. This experiment will focus on the frequency range 26.1–27.1 GHz, in order to undertake a direct test of the Beck result to the best sensitivity achievable with the technology available to us at the beginning of the tuning run. The projected exclusion limits for this search are shown in Fig. 14.2.

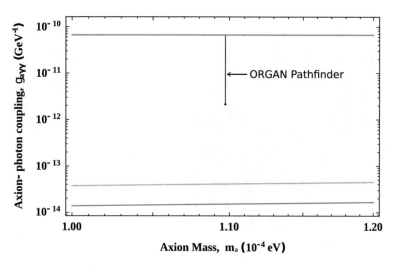

Fig. 14.1 The narrow exclusion limits on axions for the ORGAN stationary cavity pathfinding experiment. The axion KSVZ and DSFZ model bands, as well as exclusion limits from the CERN Axion Solar Telescope (CAST) are shown

14.2.2.2   Stage II

During the operation of Stage I the ORGAN collaboration will be prototyping and developing for Stage II, a wider scan from 15–50 GHz over the next 6 years, from 2019–2025. The scan will be undertaken in 5 GHz regions, Stage II-A through Stage II-G. Each region will receive approximately 10 months of data collection, during which time further research and development for the remaining future stages can be undertaken. The aims of this research will be to develop resonators at the requisite frequencies with the best possible sensitivity based on a C^2V^2G figure of merit, to develop quantum limited amplification for each frequency range, as well as a potential further magnet upgrade to 28 T. It is our goal to install quantum limited amplification at the beginning of Stage II-A at 15 GHz, and develop new amplification as we progress. The sensitivities for Stages II-A through II-G are shown in Fig. 14.2. The caption details the different scenarios associated with each set of exclusion limits. We have elected to begin our search frequency range at 15 GHz, as the region below this will be capably covered by other experiments such as those underway at ADMX, CULTASK, and Yale, and future experiments such as ABRACADABRA [9]. Regions above 50 GHz are not as well motivated theoretically.

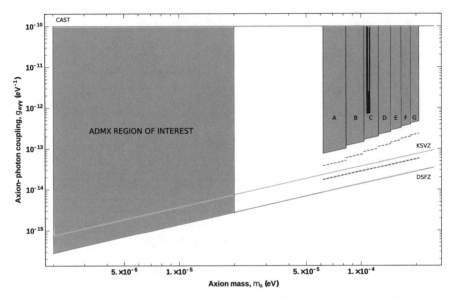

Fig. 14.2 Strength of axion to two photon coupling as a function of axion mass. The narrow navy bar represents the first stage of the experiment. The narrower aqua bar set within the navy bar represents the limits from the pathfinding experiment. A–G represent the various phases of the second stage of the experiment over the next 6 years of EQuS, the 15–50 GHz scan is broken into 5 GHz regions, assuming that the goal of quantum limited amplification is met. The dashed gray lines show the limits that could be reached with a larger 28 T magnet. The dashed red line shows the projected sensitivity under the assumption that we can surpass the quantum noise limit and achieve added noise at the level of the thermal noise of the resonator and amplifier, whereas the dashed green line represents the same, for the larger 28 T magnet

14.3 Dielectric Resonators

Much current research is focused on dielectric resonators and their application to axion detection. For the ORGAN experiment, we are developing resonators based on dielectric rings placed inside cylindrical cavities to serve as a virtual boundary condition, thus boosting quality factors via the Bragg effect [10]. Quality factor improvements on the order of five times can be achieved when compared with an empty resonator of the same dimensions. In addition to the traditional Bragg effect, it is possible to design these dielectric rings to be in the right location and of the right thickness such that the axion sensitivity of a TM mode is increased. See Fig. 14.3 for an illustration of these effects. The principle of this is as follows: it can be shown that the haloscope form factor is dependent on the relative dielectric constant of the medium. Specifically, the coupling is suppressed by a factor of ϵ_r. As such, it is possible to position the ring inside, for example, a TM_{030} mode such that the out of phase (negative) E_z field region is contained within the dielectric medium,

Fig. 14.3 z-component of electric field for two axisymmetric modes in a cylindrical resonator. The mode on the left is a TM_{020} mode confined by the thin dielectric ring, i.e. the Bragg effect. The mode on the right is a TM_{030} mode with the dielectric ring placed to suppress the out of phase z-direction electric field, and increase axion form factor

thus reducing the contribution of the out of phase field to the form factor. With careful resonator design it is possible to engineer modes based on this principle with enhanced quality factor and form factor concurrently. The optimal parameters for quality factor and form factor do not coincide, and thus a trade-off is required. Resonators based on this principle are the subject of a forthcoming publication, and it is likely that such a scheme will be implemented in ORGAN.

14.4 Cross-Correlation

Power combining a number of small cavities is a common goal of high frequency haloscopes, however, this presents a series of immense engineering challenges in practice. We present a cross-correlation scheme for effectively power-combining cavities. The principle advantages of this scheme over a traditional Wilkinson power combining scheme are that the cavities can be spatially well separated, even inside separate magnet and cryogenic systems. Furthermore, we do not need to consider the relative phase of the signals being combined to optimize sensitivity. A schematic of the cross-correlation scheme for two cavities is given in Fig. 14.4. Signals uncorrelated between channels, such as the thermal noise of the amplifier and resonators, are rejected, whereas signals that are correlated, such as any WISP signal present in the system, will be retained provided that the cavities are not separated by a distance greater than the coherence length of the particle. This opens up the possibility of characterizing the coherence length of any signal that is detected. Furthermore, it can be shown [11] that for two cavities, the sensitivity to axions of the cross-spectrum is slightly reduced compared with a traditional Wilkinson sum of two cavities, but for larger number of cavities, it is possible to post-process the data in such a way that an improvement can be attained. Specifically, if we

Fig. 14.4 A schematic of the cross-correlation scheme for two cavities, used in a proof of concept experiment

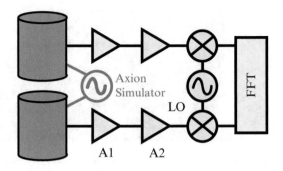

have n cavities, we can compute $\frac{n(n-1)}{2}$ independent cross-spectra, which can then be linearly averaged. This technique can yield a slight improvement over power combining n cavities traditionally, however, we require n independent measurement channels.

Acknowledgements The authors would like to thank Stephen Parker, Eugene Ivanov, Justin Kruger, Graeme Flower, Maxim Goryachev and Jeremy Bourhill.

References

1. B.T. McAllister, G. Flower, E.N. Ivanov, M. Goryachev, J. Bourhill, M.E. Tobar, The ORGAN experiment: an axion haloscope above 15 GHz. Phys. Dark Universe **18**, 67–72 (2017)
2. T.M. Shokair, J. Root, K.A. Van Bibber, B. Brubaker, Y.V. Gurevich, S.B. Cahn, S.K. Lamoreaux, M.A. Anil, K.W. Lehnert, B.K. Mitchell, A. Reed, G. Carosi, Future directions in the microwave cavity search for dark matter axions. Int. J. Mod. Phys. A **29**(19), 1443004 (2014)
3. C Woohyun, CULTASK, The Coldest Axion Experiment at CAPP/IBS in Korea, PoS, CORFU2015, 047 (2016)
4. I. Stern, A.A. Chisholm, J. Hoskins, P. Sikivie, N.S. Sullivan, D.B. Tanner, G. Carosi, K. van Bibber, Cavity design for high-frequency axion dark matter detectors. Rev. Sci. Instrum. **86**(12), 123305 (2015)
5. B.M. Brubaker, L. Zhong, Y.V. Gurevich, S.B. Cahn, S.K. Lamoreaux, M. Simanovskaia, J.R. Root, S.M. Lewis, S. Al Kenany, K.M. Backes, I. Urdinaran, N.M. Rapidis, T.M. Shokair, K.A. van Bibber, D.A. Palken, M. Malnou, W.F. Kindel, M.A. Anil, K.W. Lehnert, G. Carosi, First results from a microwave cavity axion search at 24 μ eV. Phys. Rev. Lett. **118**, 061302 (2017)
6. G. Ballesteros, J. Redondo, A. Ringwald, C. Tamarit, Unifying inflation with the axion, dark matter, baryogenesis and the seesaw mechanism. Phys. Rev. Lett. **118**, 071802 (2017)
7. C. Beck, Possible resonance effect of axionic dark matter in Josephson junctions. Phys. Rev. Lett. **111**, 231801 (2013)

8. E. Berkowitz, M.I. Buchoff, E. Rinaldi, Lattice QCD input for axion cosmology. Phys. Rev. D **92**, 034507 (2015)
9. Y. Kahn, B.R. Safdi, J. Thaler, Broadband and resonant approaches to axion dark matter detection. Phys. Rev. Lett. **117**(14), 141801 (2016)
10. M.E. Tobar, J.G. Hartnett, J.M. le Floch, D. Cros, Cylindrical distributed Bragg reflector resonators with extremely high Q-factors, in *Proceedings of the 2004 IEEE International Frequency Control Symposium and Exposition, 2004*, pp. 257–265 (2004)
11. B. McAllister, S.R. Parker, E.N. Ivanov, M.E. Tobar, Cross-correlation measurement techniques for cavity-based axion and weakly interacting slim particle searches, arXiv:1510.05775 [physics.ins-det]

Chapter 15
Searching for Low Mass Axions with an LC Circuit

N. Crisosto, P. Sikivie, N. S. Sullivan, and D. B. Tanner

Abstract Axions are a promising cold dark matter candidate. Haloscopes, which use the conversion of axions to photons in the presence of a magnetic field to detect axions, are the basis of microwave cavity searches such as ADMX. To search for lighter, low frequency axions in the unexplored $<2 \times 10^{-7}$ eV (50 MHz) range a tunable LC circuit has been proposed. Progress in the development of such an LC circuit based search is presented here. This will include preliminary results from prototypes using electrical tuning, superconducting loop antenna, and aluminum shielding.

Keywords Axion · Dark-matter · LC-circuit · Low-frequency · Coupling · SQUID · Varactor diode · Florida · SQUID · Superconducting loop

15.1 Background

The constituents of dark matter are yet to be accounted for. Axions are an exceptional dark matter candidate as they arise independently from the Peccei-Quinn solution to the strong CP Problem [1, 2]. The Sikivie haloscope detection scheme [3, 4] is based on the electromagnetic coupling of axions:

$$\mathscr{L}_{a\gamma\gamma} = -ga\mathbf{E} \cdot \mathbf{B} \tag{15.1}$$

where g is coupling constant, a is the axion field, $\mathbf{E}$ is the electric field, and $\mathbf{B}$ is the magnetic field. In the presence of a strong magnetic field, an axion may convert into a real, detectable photon. A tuned resonator can subsequently enhance detection of the axion sourced photon signal. The axion mass is a free parameter and the detector must be tuned through the range of possible frequencies corresponding

N. Crisosto (✉) · P. Sikivie · N. S. Sullivan · D. B. Tanner
Department of Physics, University of Florida, Gainesville, FL, USA
e-mail: ncrisosto@ufl.edu; sikivie@phys.ufl.edu; sullivan@phys.ufl.edu; tanner@phys.ufl.edu

© Springer International Publishing AG, part of Springer Nature 2018   127
G. Carosi et al. (eds.), *Microwave Cavities and Detectors for Axion Research*,
Springer Proceedings in Physics 211, https://doi.org/10.1007/978-3-319-92726-8_15

to the axion mass. Frequency of the axion sourced photon signal, ω, is set by the condition $\hbar\omega = m_a c^2$, where m_a is the axion mass. Microwave cavity searches, including ADMX and HAYSTAC, have already scanned sections of axion parameter space [5–10].

To look for lighter, low frequency axions an LC circuit could be used instead of a microwave cavity [11]. Equation (15.1) implies that an oscillating axion field in a static externally applied magnetic field $\mathbf{B}_0(\mathbf{x})$, from a solenoid magnet in this example, produces a current density along the $\mathbf{B}_0$ magnetic field lines; see Eq. (15.2) below. From Ampere's law there is a subsequent $\mathbf{B}_a$ field oscillating around the axis of the externally applied magnetic field. A loop antenna inductor can capture this magnetic flux. When the LC circuit is tuned on resonance with $\mathbf{B}_a$ the captured flux drives a current in the LC circuit. See Fig. 15.1. The current in the LC circuit can be coupled into a SQUID for amplification. A variable capacitor can be used to tune the LC circuit resonance through a range of possible axion masses.

Outlining the detection principle in more detail, from Eq. (15.2), in an externally applied magnetic field, $\mathbf{B}_0$, there is an effective exotic axion based current

$$\mathbf{j}_a = -g\mathbf{B}_0 \frac{\partial a}{\partial t} \tag{15.2}$$

and subsequently

$$\nabla \times \mathbf{B}_a = \mathbf{j}_a. \tag{15.3}$$

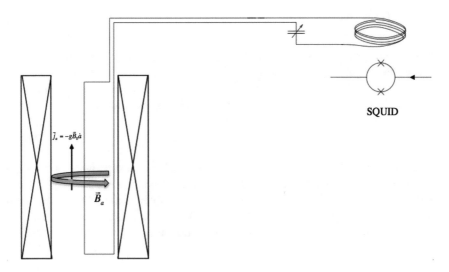

Fig. 15.1 Diagram of an LC circuit axion search, using an externally applied magnetic field from a solenoid, adapted from [11]

From $\mathbf{B_a}$, a loop, with inductance L, inserted into the solenoid magnet space captures a flux Φ_a,

$$\Phi_a = -V_m g \frac{\partial a}{\partial t} B_0. \tag{15.4}$$

Using the definition of inductance it follows that

$$I = -\frac{\Phi_a}{L}. \tag{15.5}$$

Along with resonant enhancement this results in

$$I = \frac{Q}{L} V_m g \frac{\partial a}{\partial t} B_0 \tag{15.6}$$

where $V_m = \frac{1}{4} l_m r_m^2$, l_m is the length, r_m the width of the primary loop, and Q is the quality factor of the LC circuit when tuned on resonance with $\mathbf{B_a}$. Note that the size of the primary loop is only half of the magnet bore diameter. A loop spanning the entire diameter would capture zero net magnetic flux. The current given in Eq. (15.6) could then be read out to a SQUID. In the initial pilot run a GaAs FET will be used.

15.2 Prototype LC Circuit Testing

Critical components of an effective resonator design are a high quality factor (Q) and the ability to tune through a wide frequency range. Various LC circuit designs have been built and compared based on Q and tuning range as performance metrics. Iterative changes have been made in various prototypes to understand loss mechanisms. A tracking generator from a Rigol Spectrum Analyzer is used to excite the LC circuit and the resonant response is read out by an additional weakly coupled probe; see also Fig. 15.2.

15.2.1 Room Temperature and 77 K Testing

Loop antennas were first made out of oxygen free high thermal conductivity copper (OFHC) and tuned with air gap mechanical capacitors. Quality factors of 300 in open air, 600 with shielding, and 1200 with liquid nitrogen cooling and shielding were achieved.

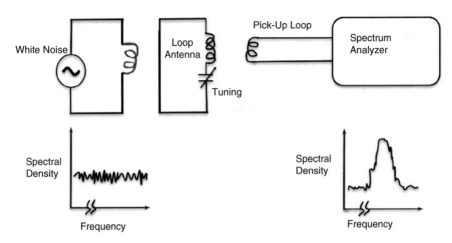

Fig. 15.2 Diagram of the frequency-response measurement method

15.2.2   4.2 K Testing

Cooling to cryogenic temperatures was done to reduce resistive and dielectric losses of the LC circuit, as well as make the use of superconducting materials viable. A liquid helium dewar was setup with two stainless steel sma coaxial cable lines for frequency response measurements. The dewar top plate was electrically floated with a Delrin® ring and o-rings on both sides.

15.2.2.1   Cooper Loop with Varactor Tuning

Varactors were considered as a possible tuning mechanism of the LC circuit. Varactor capacitance changes with the applied bias voltage enabling electrical tuning. Incorporating varactor tuning, with a GaAs MACOM MA46H202-1056 diode, into a copper loop antenna worsened Q at room temperature to below 300 throughout the tuning range. Cooling the varactor tuned cooper loop further reduced Q below 100 as shown in Fig. 15.3. The Q reduction is attributed to an increased loss in the varactor at low temperatures.

15.2.2.2   Superconducting Loop with Parallel Plate Capacitor

A single turn loop of copper-clad NbTi wire was strung on a polyether ether ketone (PEEK) form and silver-soldered to OFHC parallel plates that formed a capacitor. The capacitive plates were held in place by insulating PEEK screws. Upon cooling, a Q of 1300 was attained at 38.3 MHz, see Fig. 15.4. This is the highest Q ever reached, but is comparable to OFHC loop performance at 77 K, which suggests

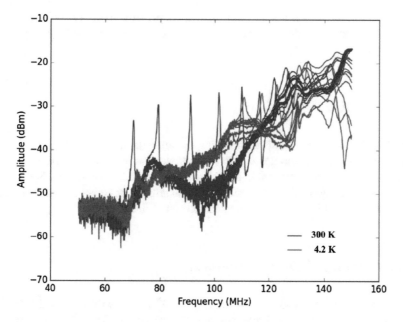

Fig. 15.3 LC circuit resonant frequency varying with changing varactor bias voltage from 0–12 V. Full resonant frequency ranges at 300 and 4.2 K are shown for comparison

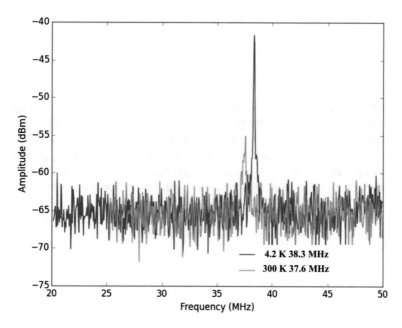

Fig. 15.4 Results from a liquid helium cooled superconducting loop antenna

electrical conductivity is not the limiting factor at present. To limit possible radiation losses aluminum shields were added for a later cool down, but without significant Q change.

15.2.3   Plans Forward

The shields will be made superconducting by lining the interior with lead to further reduce possible radiation losses. If Q improvement is found, a magnetic field compatible superconducting shield, such as NbTi will be incorporated. Mechanical tuning by inserting sapphire between existing parallel plates gives a projected tuning range of 18–38 MHz and is the intended method of future tuning.

15.2.4   Pilot LC-Circuit Experiment

For the pilot low-frequency axion search, the LC circuit will be cooled by a ^{3}He refrigerator with estimated 1 mW cooling power at 0.4 K. The LC Circuit will be used in a magnetic field from a NbTi solenoid magnet with 8.6 T center field, 70 cm long, and 17 cm bore. A GaAs cryogenic FET amplifier will be used to amplify the axion field signal prior to detection.

15.3   Conclusion

A LC circuit based axion search presents a promising approach to scan unexplored regions of low-frequency axion parameter space. LC circuits built thus far have suggested that varactor tuning is too lossy and that fixed capacitor loop antenna Q factors are limited by losses other than electrical conductivity. On going prototyping will continue with the addition of superconducting shields and mechanical tuning. The culmination of this prototyping will be a pilot experiment exploring low-mass axions.

Acknowledgements This research was supported by DOE grants DE-SC0010280 and DE-SC0010296.

References

1. R.D. Peccei, H.R. Quinn, Phys. Rev. Lett. **38**, 1440 (1977)
2. S. Weinberg, Phys. Rev. Lett. **40**, 223 (1978)

3. P. Sikivie, Phys. Rev. Lett. **51**, 1415 (1983)
4. P. Sikivie, Phys. Rev. D **32**, 2988 (1985)
5. S. DePanfilis et al., Phys. Rev. Lett. **59**, 839 (1987)
6. C. Hagmann, P. Sikivie, N.S. Sullivan, D.B. Tanner, Phys. Rev. D **42**, 1297 (1990)
7. S. Asztalos et al., Phys. Rev. Lett. **104**, 041301 (2010)
8. J.V. Sloan et al., Phys. Dark Univ. **14**, 95 (2016)
9. J. Hoskins et al., Phys. Rev. D **94**, 082001 (2016)
10. B.M. Brubaker et al., Phys. Rev. Lett. **118**, 061302 (2017)
11. P. Sikivie, N. Sullivan, D.B. Tanner, Phys. Rev. Lett. **112**, 131301 (2014)

Chapter 16
ABRACADABRA: A Broadband/ Resonant Search for Axions

Yonatan Kahn for the ABRACADABRA Collaboration

Abstract In the long-wavelength regime, axion interactions with a static magnetic field can be described in terms of an effective current which sources a small oscillating magnetic field. I will describe a new experiment (ABRACADABRA) to detect this axion effective current which can operate with either broadband or resonant readout of the signal. Inspired by advances in medical physics and precision magnetometry, the broadband approach has advantages at low axion mass and can probe many decades of mass simultaneously. The combination of broadband and resonant approaches potentially has sensitivity to GUT-scale QCD axions. I will discuss recent progress on a prototype of ABRACADABRA under development at MIT.

Keywords Axion · Cold dark matter · ABRACADABRA · Broadband · Resonant · MHz · kHz · Effective current · Toroidal magnet · Superconducting loop · SQUID

ABRACADABRA Collaboration: Lindley Winslow (PI), MIT; Janet Conrad, Joseph Formaggio, Sarah Heine, Joe Minervini, Jonathan Ouellet, Kerstin Perez, Nicholas Rodd, Alexey Radovinsky, Jesse Thaler, Daniel Winklehner, MIT; Reyco Henning, University of North Carolina at Chapel Hill; Joshua Foster, Benjamin R. Safdi, University of Michigan at Ann Arbor; Yonatan Kahn, Princeton University.

Y. Kahn (✉)
Princeton University, Princeton, NJ, USA
e-mail: ykahn@princeton.edu

© Springer International Publishing AG, part of Springer Nature 2018   135
G. Carosi et al. (eds.), *Microwave Cavities and Detectors for Axion Research*,
Springer Proceedings in Physics 211, https://doi.org/10.1007/978-3-319-92726-8_16

16.1 Axion Dark Matter

Axions are an excellent dark matter (DM) candidate, with the added benefit of solving the strong-CP problem. While much attention has been given to axions in the GHz frequency range, lower-frequency axions (kHz-MHz) are also well-motivated by UV physics considerations. If the Peccei-Quinn (PQ) symmetry for which the axion is a pseudo-Goldstone boson is broken before the end of inflation, the cosmic abundance of axions is set by the misalignment mechanism (see [1] for a review), and the observed relic density is given by

$$\Omega h^2 \sim 0.1 \left(\frac{f_a}{10^{16} \text{ GeV}} \right)^{7/6} \left(\frac{\theta_i}{5 \times 10^{-3}} \right)^2, \tag{16.1}$$

where f_a is the PQ breaking scale (also known as the axion decay constant) and θ_i is the initial misalignment angle. We see that axions with GUT-scale decay constants can simultaneously explain dark matter and solve the strong-CP problem, at the price of a mild $\sim 1\%$ tuning. The corresponding axion mass is

$$m_a \sim 6 \times 10^{-10} \text{ eV} \left(\frac{10^{16} \text{ GeV}}{f_a} \right), \tag{16.2}$$

which as we will describe below, gives rise to oscillating EM signals at ~ 100 kHz.

16.1.1 Axion Field Properties

Axion DM today behaves like a classical field [2]: for $m_a \lesssim 1$ eV, the observed local density of DM implies macroscopic occupation numbers. The axion field dynamics are determined by the Klein-Gordon equation, with spatially homogeneous solution

$$a(t) = a_0 \sin(m_a t) = \frac{\sqrt{2\rho_{\text{DM}}}}{m_a} \sin(m_a t). \tag{16.3}$$

Spatial gradients of the axion field are suppressed by the virial velocity of DM, $v_{\text{DM}} \sim 10^{-3} c$. This small virial velocity, and correspondingly small velocity dispersion, also leads to spatial and temporal phase coherence of the axion field over macroscopic distances and times:

$$\lambda_{\text{coh}} \sim \frac{2\pi}{m_a v_{\text{DM}}} \approx 100 \text{ km} \frac{10^{-8} \text{ eV}}{m_a} \tag{16.4}$$

$$\tau_{\text{coh}} \sim \sim \frac{2\pi}{m_a v_{\text{DM}}^2} \approx 0.4 \text{ s} \frac{10^{-8} \text{ eV}}{m_a} \tag{16.5}$$

This is highly advantageous for terrestrial axion experiments, as signal-to-noise can be improved by taking data over many phase-coherent periods.

16.1.2 Axion Effective Current

Axions, or axion-like particles, will generically have an interaction with photons of the form

$$\mathcal{L} \supset -\frac{1}{4} g_{a\gamma\gamma} F_{\mu\nu} \widetilde{F}^{\mu\nu}. \tag{16.6}$$

For the QCD axion, $g_{a\gamma\gamma} \sim \frac{\alpha}{\pi} \frac{1}{f_a} \propto m_a$ up to order-1 model uncertainties, but for a general axion-like particle unrelated to the strong-CP problem, $g_{a\gamma\gamma}$ and m_a are independent model parameters. This interaction leads to additional terms in Maxwell's equations [3]. In the presence of a background axion field $a(t)$ and a static background magnetic field $\mathbf{B}_0$, the leading effect is

$$\nabla \times \mathbf{B} = -\frac{\partial \mathbf{E}}{\partial t} + g_{a\gamma\gamma} \mathbf{B}_0 \frac{\partial a}{\partial t}. \tag{16.7}$$

We can read the right-hand side as a source term in Ampère's Law, and thus as an effective current density

$$\mathbf{J}_{\text{eff}} = g_{a\gamma\gamma} \sqrt{2\rho_{\text{DM}}} \cos(m_a t) \mathbf{B}_0. \tag{16.8}$$

Note that this current follows lines of the external field $\mathbf{B}_0$, and is an AC response to a DC external field, with frequency set by the axion mass.

16.2 ABRACADABRA: Theory

The ABRACADABRA (A Broadband/Resonant Approach to Cosmic Axion Detection with an Amplifying B-field Ring Apparatus) proposal [4] aims to detect the axion-induced effective current, Eq. (16.8), in the magnetoquasistatic (MQS) regime where λ_a, the axion Compton wavelength, is much larger than the size of the experiment. In this regime, the relevant observable is the local axion-induced magnetic flux, rather than the radiated power at infinity from the oscillating current source.

Fig. 16.1 Schematic of the
ABRACADABRA toroid

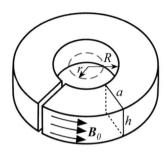

16.2.1 Toroidal Geometry

ABRACADABRA utilizes a toroidal magnetic field, shown in Fig. 16.1. The
effective current $\mathbf{J}_{\mathrm{eff}}$ follows the lines of $\mathbf{B}_0$ which circulate around the toroid,
and sources an AC magnetic flux which goes through the center hole of the toroid.
This flux is detected by a superconducting pickup loop, shown in dashed red. The
advantage of a toroidal geometry is that the region where the flux is detected is
outside of the strong-field regime. In the absence of an axion, the only fields in the
center of the toroid are fringe fields. In an ideal setup where the field is sourced
by a superconducting current, the fringe fields are pure DC; vibrational noise will
likely give a small AC component, but one which is much smaller than a geometry
where the pickup is in the strong-field region, as in the proposal of [5]. The toroidal
geometry is also reminiscent of cryogenic current comparators (CCC) [6], which are
used for precision current metrology at low frequencies; here $\mathbf{J}_{\mathrm{eff}}$ plays an analogous
role to a real current, which we can exploit to inject test signals in an experimental
realization. The gap in the toroid (often an overlap in CCC applications) is to ensure
the signal does not get screened by Meissner currents at sufficiently low frequencies.
The signal flux is

$$\Phi_a(t) = g_{a\gamma\gamma} \sqrt{2\rho_{\mathrm{DM}}} \cos(m_a t) \times (B_{\max} V G_{\mathrm{toroid}}), \tag{16.9}$$

where $B_{\max}$ is the maximum B-field strength in the toroid, V is the toroid volume,
and G_{toroid} is a geometric factor which is typically $\mathcal{O}(0.1)$.

16.2.2 Broadband and Resonant Readout

Another advantage of the ABRACADABRA design is that it does not rely on
a resonant enhancement to see a measurable signal. The circuit containing the
pickup loop can either be a broadband (i.e. purely inductive) circuit, or an LC
resonant circuit containing a capacitor and tuned to a particular frequency. In
either case, the pickup circuit is inductively coupled to a SQUID magnetometer.
Studies of weak-field MRI in medical physics have shown that broadband (or

untuned) pickup circuits coupled to SQUIDS can have superior signal-to-noise at low frequencies, depending on the Q-factor of the capacitor [7]. In the broadband circuit, which cannot resolve the irreducible thermal noise, the dominant irreducible source of noise is SQUID flux noise, which can be as low as $10^{-6}\Phi_0/\sqrt{Hz} = 2.1 \times 10^{-21}$ Wb/$\sqrt{Hz}$. Taking data for a time $t \gg \tau_{coh}$, a signal-to-noise ratio of 1 can be achieved for couplings of

$$
g_{a\gamma\gamma} > 6.3 \times 10^{-18} \text{ GeV}^{-1} \left(\frac{m_a}{10^{-12} \text{ eV}} \frac{1 \text{ year}}{t} \right)^{1/4} \frac{5 \text{ T}}{B_{max}}
$$
$$
\times \left(\frac{0.85 \text{ m}}{R} \right)^{5/2} \sqrt{\frac{0.3 \text{ GeV/cm}^3}{\rho_{DM}}} \frac{S_{\Phi,0}^{1/2}}{10^{-6}\Phi_0/\sqrt{Hz}}. \qquad (16.10)
$$

This is already sufficient to probe the GUT-scale axion of Eq. (16.2).

16.2.3   Projected Reach

Using a combination of broadband and resonant readout circuits, a full-scale ABRACADABRA experiment with $VG_{toroid} = 100 \text{ m}^3$ can probe the QCD axion with a GUT-scale decay constant; see Fig. 16.2. The resonant curves assume a particular scanning strategy where each e-fold of frequency shown on the plot is covered in equal time, for a total data-taking time of 1 year, and for which data is only taken on the resonance peak. The Q-factor is assumed to be 10^6 at a temperature of 0.1 K. The broadband circuit also assumes a total data-taking time of 1 year. In an ideal experiment, the resonant curves can surpass the broadband curves by using a hybrid strategy which takes data at all frequencies [11].

16.3   ABRACADABRA: Experiment

A small-scale prototype, ABRACADABRA-10cm (shown in Fig. 16.3), is currently being developed at MIT. The experiment is designed to fit inside the Oxford Instruments Triton 400 dilution refrigerator already at MIT, with a 12 L working volume and a minimum temperature of 10 mK. ABRACADABRA-10cm has an inner toroid radius $R_{in} = 3$ cm, an outer radius of $R_{out} = 6$ cm, and height $h = 12$ cm for a volume of 680 cm^3. The maximum B-field is 1 T, and the geometric factor for this design is $G = 0.085$. A signal can be injected by running current through a calibration loop threading the toroid. Fringe fields in the center should be at the level of $10^{-6} B_{max}$. A Magnicon SQUID with typical flux noise of $1.2 \times 10^{-6}\Phi_0/\sqrt{Hz}$ at 4 K will be used for the pickup circuit. Magnetic shielding surrounding the experiment will be required; we are currently

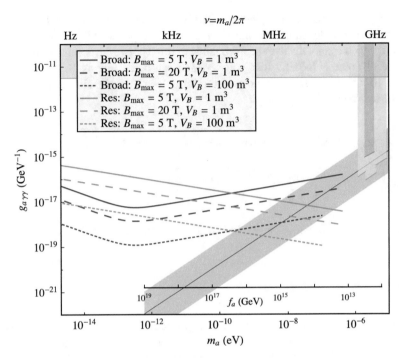

Fig. 16.2 Projected reach for ABRACADABRA for various toroid sizes, field strengths, and readout circuits. Here $V_B = V\, G_{\text{toroid}}$. The QCD axion parameter space is shown in red, exclusions from ADMX [8] and are in grey, and projections from IAXO [9] and ADMX [10] are shown in green

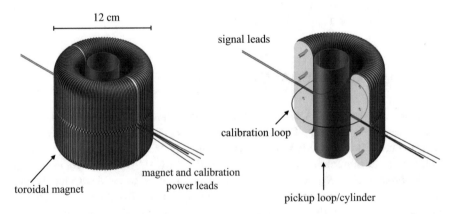

Fig. 16.3 Schematic of ABRA-10cm prototype

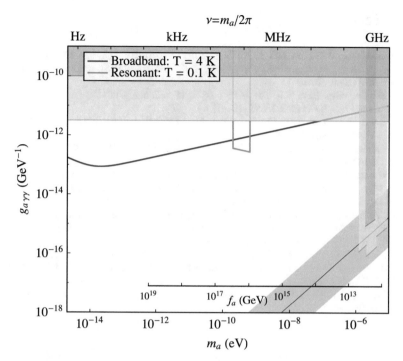

Fig. 16.4 Reach curves for ABRACADABRA-10cm for 1 month of data taking. The blue curve is for a broadband circuit at $T = 4$ K, and the orange curve is for a resonant circuit at $T = 0.1$ K scanning frequencies near 100 kHz

investigating superconducting shield designs. We expect vibrational noise to be sourced by fringe fields and to scale as $S_{\Phi,\text{vib}}^{1/2} \sim (10^{-6} B_{\text{max}}) R_{\text{out}}$, which is already subdominant compared to SQUID noise at frequencies above 1 kHz.

ABRACADABRA-10cm will take data in June 2018. Figure 16.4 shows the broadband and resonant reach for 1 month of data-taking, which will already give world-leading limits on the axion-photon coupling $g_{a\gamma\gamma}$ at Hz-MHz frequencies.

Acknowledgements ABRACADABRA-10cm is funded by NSF-1658693. The Conrad group's contribution is funded through NSF-1505858. Jesse Thaler is supported by DOE grant contract numbers DE-SC-00012567 and DE-SC-00015476.

References

1. D.J.E. Marsh, Phys. Rep. **643**, 1 (2016). https://doi.org/10.1016/j.physrep.2016.06.005
2. P.W. Graham, S. Rajendran, Phys. Rev. **D88**, 035023 (2013). https://doi.org/10.1103/PhysRevD.88.035023
3. P. Sikivie, Phys. Rev. Lett. **51**, 1415 (1983). https://doi.org/10.1103/PhysRevLett.51.1415 [Erratum: Phys. Rev. Lett. **52**, 695 (1984)]
4. Y. Kahn, B.R. Safdi, J. Thaler, Phys. Rev. Lett. **117**(14), 141801 (2016). https://doi.org/10.1103/PhysRevLett.117.141801
5. P. Sikivie, N. Sullivan, D.B. Tanner, Phys. Rev. Lett. **112**(13), 131301 (2014). https://doi.org/10.1103/PhysRevLett.112.131301
6. K. Grohmann, H. Hahlbohm, H. Lübbig, H. Ramin, Cryogenics **14**(9), 499 (1974)
7. W. Myers, D. Slichter, M. Hatridge, S. Busch, M. Mößle, R. McDermott, A. Trabesinger, J. Clarke, J. Magn. Reson. **186**(2), 182 (2007)
8. S.J. Asztalos, G. Carosi, C. Hagmann, D. Kinion, K. van Bibber, M. Hotz, L.J. Rosenberg, G. Rybka, J. Hoskins, J. Hwang, P. Sikivie, D.B. Tanner, R. Bradley, J. Clarke, Phys. Rev. Lett. **104**, 041301 (2010). https://doi.org/10.1103/PhysRevLett.104.041301
9. J.K. Vogel et al. (2013). http://inspirehep.net/record/1219323/files/arXiv:1302.3273.pdf
10. T.M. Shokair et al., Int. J. Mod. Phys. **A29**, 1443004 (2014). https://doi.org/10.1142/S0217751X14430040
11. S. Chaudhuri, K. Irwin, P. Graham, J. Mardon, arXiv:1803.01627. http://inspirehep.net/record/1658641?ln=en

Chapter 17
Searching Axions through Coupling with Spin: The QUAX Experiment

N. Crescini, D. Alesini, C. Braggio, G. Carugno, D. Di Gioacchino, C. S. Gallo, U. Gambardella, C. Gatti, G. Iannone, G. Lamanna, A. Lombardi, A. Ortolan, R. Pengo, G. Ruoso, and C. C. Speake

Abstract The axion is one of the main candidates of dark matter and was originally introduced to solve the strong CP problem of quantum chromodynamics. Here we present a proposal to search for QCD axions with mass in the $200\,\mu\mathrm{eV}$ range, assuming that they make a dominant component of dark matter. Since the axion can couple to the spin of a fermion, its presence can be seen as the existence of an equivalent rf field with frequency and amplitude fixed by the axion mass and coupling, respectively. This equivalent magnetic field would produce spin flips in a magnetic sample placed inside a static magnetic field, which determines the resonant interaction at the Larmor frequency. Spin flips would subsequently emit radio frequency photons that can be detected by a suitable quantum counter in an

N. Crescini (✉)
Department of Physics and Astronomy Galileo Galilei, University of Padova, Padova, Italy

INFN, Laboratori Nazionali di Legnaro, Legnaro, Italy
e-mail: nicolo.crescini@phd.unipd.it

D. Alesini · D. Di Gioacchino · C. Gatti
INFN, Laboratori Nazionali di Frascati, Frascati, Roma, Italy

C. Braggio · G. Carugno
Dipartimento di Fisica e Astronomia, Padova, Italy

INFN, Sezione di Padova and Dipartimento di Fisica e Astronomia, Padova, Italy

C. S. Gallo
Dipartimento di Fisica e Astronomia, Padova, Italy

INFN, Sezione di Padova, Padova, Italy

U. Gambardella · G. Iannone
INFN, Sezione di Napoli and University of Salerno, Fisciano, Italy

G. Lamanna
INFN, Sezione di Pisa and University of Pisa, Pisa, Italy

A. Lombardi · A. Ortolan · R. Pengo · G. Ruoso
INFN, Laboratori Nazionali di Legnaro, Legnaro, Italy

C. C. Speake
School of Physics and Astronomy, University of Birmingham, West Midlands, UK

© Springer International Publishing AG, part of Springer Nature 2018
G. Carosi et al. (eds.), *Microwave Cavities and Detectors for Axion Research*,
Springer Proceedings in Physics 211, https://doi.org/10.1007/978-3-319-92726-8_17

ultra-cryogenic environment. An updated report of the experimental results will be presented, together with a preliminary measurement and a projection of future improvements.

Keywords Axion · Dark matter · Wind · Spin · Coupling · Larmor frequency · Cavity · Hybridization · YIG

17.1 Introduction

The latest result reported by Planck shows that in our universe the dark matter fraction is $\Omega_{dm} \simeq 0.258$ [1]. However, the nature of such component is still unknown, apart its gravitational interaction with ordinary baryonic matter. In 1977 a new particle was introduced by Peccei and Quinn to solve the strong CP problem: the axion [2], which emerges naturally as a favored candidate for dark matter. Axions have a mass m_a inversely proportional to the energy scale f_a, at which the PQ symmetry breaks. For certain ranges of f_a and m_a (typically with masses ranging from μeV to meV), large quantities of axions are allowed to be produced in the early Universe that could account for a portion or even the totality of cold dark matter. The QUAX (QUaerere AXion) [3, 4] proposal explores in details the ideas of [5–9], which present different studies on the interaction of the cosmological axion with the spin of a fermion. Hereafter we consider the case of the interaction between cosmological DFSZ axions and the electronic spin. Since Earth is immersed in the dark matter halo of the Milky Way, it is effectively moving with the Solar System through the cold dark matter cloud, and an observer on Earth will see such axions as a wind. In particular, the effect of the axion wind on a magnetized material can be described as an equivalent oscillating rf field with frequency determined by m_a. Thus, a possible detector for the axion wind can be a magnetized sample with Larmor resonance frequency tuned to the axion mass by means of an external polarizing static magnetic field B_0 (e.g. $B_0 \sim 1.7\,T$ for 48 GHz, corresponding to a 200 μeV axion mass). The interaction with the axion effective field will drive the total magnetization of the sample, and thus produce oscillations in the magnetization that, in principle, can be detected. The sample is placed inside a microwave cavity, which limits the phase space in order to avoid the radiation damping issue and optimize the detection scheme. We also expect to have an annual modulation of the phase and a full daily modulation of the signal amplitude due to the motion of Earth in the Solar System, which is a strong signature of the searched signal.

17.2 Experimental Technique

Assuming the presence of a dark matter halo entirely composed by axions, the axionic wind, in the form of an effective rf magnetic field, can be sensed by a magnetized material. Its value and frequency are:

$$B_a = 2.0 \cdot 10^{-22} \left(\frac{m_a}{200 \, \mu eV} \right) \text{T}, \quad \frac{\omega_a}{2\pi} = 48 \left(\frac{m_a}{200 \, \mu eV} \right) \text{GHz}. \qquad (17.1)$$

To detect the extremely small rf field B_a we will exploit Electron Spin Resonance (ESR) in a magnetic sample. The axions deposit a small amount of power in the sample because of their interaction with the electronic spin, we want to collect this deposited power. To enhance the interaction we will tune the ESR resonance of the sample (i.e. the Larmor frequency of the electron) to the searched mass value of the axion. In fact, since this is still unknown, the possibility to perform a large band search must be envisaged. Let us consider a magnetized sample of volume V_S and magnetization M_0 placed in the bore of a solenoid, which generates the static magnetic field B_0 (polarizing field) in the z direction. The value B_0 determines the Larmor frequency of the electrons, and so the axion mass under scrutiny, through the relation

$$B_0 = \frac{\omega_L}{\gamma} = \frac{m_a c^2}{\gamma \hbar} = 1.7 \left(\frac{m_a}{200 \, \mu eV} \right) \text{T}. \qquad (17.2)$$

In the presence of the axion effective field $B_a(t)$ a time dependent component of the magnetization $M_a(t)$ will appear in the (x, y) plane. It is to be noted that only the component of the axion effective field orthogonal to the magnetizing field B_0 will drive the magnetization of the sample, thus configuring this apparatus as a true directional detector. It is

$$M_a(t) = \gamma \mu_B B_a n_S \tau_{\min} \cos(\omega_a t), \qquad (17.3)$$

where n_S is the material spin density and $\tau_{\min}$ is the shortest coherence time among the following processes: axion wind coherence $\tau_{\nabla a}$, magnetic material relaxation τ_2, radiation damping τ_r. Figure 17.1 shows the principle of the proposed detection scheme: a microwave resonant cavity, containing a magnetic material, is kept at very low temperature and placed inside an extremely uniform magnetizing field B_0, up to 1 ppm. The magnetic field value determines the Larmor frequency $\omega_m = \gamma B_0$ of the ferromagnetic resonance of the magnetic material. Using a cavity with a resonant frequency that can be reached by ω_m tuning B_0 and thus changing the ESR resonance frequency, it is possible to achieve the coupling between the electronic spin and the electromagnetic field stored in the cavity. Considering a cylindrical cavity, the sample material is placed along its z-axis, where is located the maximum of the TM110 magnetic field [see Fig. 17.1]. The single spin coupling is $g_0 = \gamma \sqrt{\mu_0 \hbar \omega_m / V_c}$ (In units of rad/s), where V_c is the volume of the cavity mode. If the total number of spin is large enough, the single resonance splits into two with a mode separation given by the total coupling strength $g_m = g_0 \sqrt{n_S V_S}$, where n_S is the spin density of the material and V_S is the volume of the material. In the presence of the axion wind, the average amount of power absorbed by the material in each cycle is

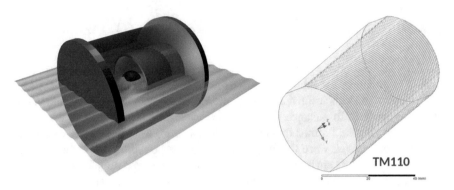

Fig. 17.1 Left: scheme of the haloscope: in green is represented the effective rf field (note that the wavelength is not to scale) and in blue the B_0 magnetic field. The right part of the cavity is reported in copper, to show the inner part, containing the black sample sphere. The whole image is not to scale with the real apparatus. Right: magnetic profile of the TM110 mode

$$P_{\text{in}} = \mu_0 \mathbf{H} \cdot \frac{\mathbf{M}}{t} = B_a \frac{M_a}{t} V_S = \gamma \mu_B n_S \omega_a B_a^2 \tau_{\text{min}} V_S$$

where we have used Eq. (17.3) for the axion induced resonant magnetization M_a. Assuming a steady state, the absorbed power can be collected using an antenna critically coupled to the interested cavity mode and is $P = P_{\text{in}}/2$ because of the power balance of the state. The limit of the sensitivity will be essentially determined by the thermal noise in the system.

17.3 First Results

Hereafter we report some preliminary tests, performed in order to verify some of the main assumptions of the scheme: the absence of extra noise added by the sample and the possibility to increase the volume of magnetic material inside the cavity.

The first is an experimental test to verify the correspondence between the thermal noise in an empty cavity or in a cavity with a magnetic sample placed inside it. Using a linear amplifier, we measured the power delivered by a microwave resonant cavity in three different conditions: (a) an empty cavity; (b) a cavity with a magnetic material inside but no magnetizing field; (c) with a magnetic material subjected to a magnetizing field such that hybridization occurs. Since the measurements (a) and (b) give the same results (apart from a small frequency shift), we report only on the comparison between (b) and (c) conditions.

For this measurement we used a copper cavity shaped in the form of a parallelepiped with dimensions along x, y, z axes 85 mm, 40 mm and 10 mm, respectively. The resonance frequency of the selected TE102 mode is $\omega_c/2\pi = 5.154$ GHz, with a total cavity linewidth $k_c/2\pi = 2.4$ MHz. The magnetic material is a sphere of YIG (Yttrium Iron Garnet, $n_S = 2 \times 10^{28}$ m^3) of 2 mm diameter placed exactly at the center of the cavity. We used a magnetizing field $B_0 = 0.19$ T, placed along the z direction of the cavity, i.e. orthogonal to the radio frequency magnetic field. Since in this conditions $\omega_m = \omega_c$, the system hybridizes. A mobile loop antenna is tuned to achieve critical coupling with the cavity mode, and measures the power delivered by the cavity. Without any input signal, the measured power is due to thermal photons stored in the resonator. The antenna output is amplified by a cascade of two cryogenic amplifiers (kept at a temperature of 77 K), the first one having a noise temperature well below the cavity temperature. The final output, amplified by about 69 dB, is then fed into a spectrum analyzer for recording. The results of these measurements are given in Fig. 17.2.

For calibration purposes a 50 Ω load was put at the input of the amplifying chain: this allows to check the gain versus frequency curve and the nominal input level. As can be seen from the figure, the cavity peak power level correspond to the Johnson noise of the 50 Ω resistor. The separation between the two peaks of the hybridized system is about 80 MHz and the expected value is $g_m/2\pi = 90$ MHz. From the fit of the data one gets $k_m/2\pi = 1.0$ MHz, which corresponds to the natural linewidth of YIG associated with the spin-spin relaxation time $\tau_2 = 0.16\,\mu$s. The height of the two hybridized modes coincides again with the Johnson noise of the 50 Ω load and consequently of the empty cavity, so we verified that at room temperature the magnetic sample does not add extra noise inside the cavity. Given this fact, we

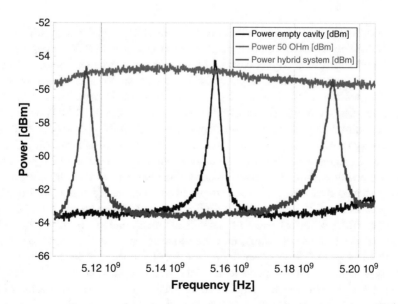

Fig. 17.2 Thermal noise measured with a linear amplifier (total gain = 69 dB, RBW = 100 kHz). $T_c = 300$ K. The antenna coupling has been optimized for each individual measurement. Red curve: amplifier input connected to a 50 Ω load. Black curve: empty cavity output. Blue curve: hybrid system output

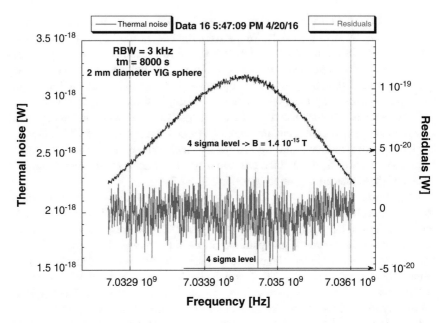

Fig. 17.3 Measurement at 7.0 GHz, corresponding to an axion mass of $m_a = 37\,\mu\text{eV}$. The black line is the resonance of the cavity, while in pink are represented the residuals. For this measurement we used a RBW of 3 kHz

proceeded with a second test, with the aim of providing a preliminary limit on the searched interaction. The measurement is reported in Fig. 17.3. We used the same 2 mm YIG sphere in a copper cavity with a mode resonating at a frequency $\omega_c = 7.0$ GHz. Tuning the external magnetic field to a value such that $\omega_c \simeq \omega_m$ we obtain hybridization and average data for one of the two hybrid resonances for 8000 s. After the integration, from the residuals we obtained $P < 5 \times 10^{-20}$ W at 4σ level, corresponding to a limit on the equivalent magnetic field $B_a < 1.4 \times 10^{-15}$ T.

Another issue is the possibility of using a large quantity of magnetic material, up to 0.1 L, to enhance signal of the axionic wind. For this reason we performed a measurement with 3 YIG spheres placed along the z axis of a cylindrical cavity, evenly spaced by 8 mm from the central sphere. The cavity was placed inside a homogeneous static magnetic field parallel to z. The three spheres have been put in place with the correct value of static field corresponding to $\omega_m = \omega_c$. After obtaining the best hybridization, they are removed one at a time and the result is reported in Fig. 17.4. It was seen that, as expected, the value of the coupling scales exactly with the square root of the total number of spins. This shows that it is feasible to use cavities with a large longitudinal dimension to increase the volume of the magnetic material: all the spins participate to the process of coupling to the cavity mode.

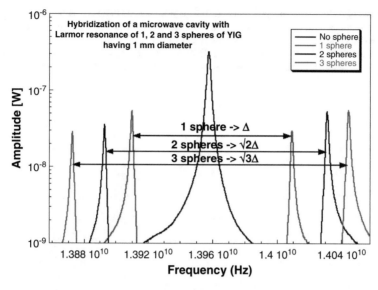

Fig. 17.4 Hybridization measurement with three, two, one and no spheres inside the cavity (see text for further details). The peaks are of different heights due to differences in antenna coupling for the various modes, since coupling is never optimized

17.4   Conclusions

This scheme aims to detect the modulation of the magnetization of a sample, due to the presence of an axionic wind acting as an equivalent rf field. The use of the Larmor resonance of a material allows enough sensitivity to search for the DFSZ axion, assuming that the local dark matter density is entirely composed by axions. Working in an ultra-cryogenic environment, and developing a microwave photon counter might allow the scanning of an interesting mass range expected for the QCD axion.

References

1. Planck Collaboration: P.A.R. Ade et al., Astron. Astrophys. **594**, A13 (2016)
2. R.D. Peccei, H.R. Quinn, Phys. Rev. Lett. **38**, 1440 (1977); Hwang, P. Sikivie, D. Tanner, R. Bradley, J. Clarke, Phys. Rev. Lett. **104**, 041301 (2010)
3. R. Barbieri et al., Phys. Dark Univ. **15**, 135141 (2017)
4. G. Ruoso et al., J. Phys. Conf. Ser. **718**, 042051 (2016)
5. L.M. Krauss, J. Moody, F. Wilczeck, D.E. Morris, Spin coupled axion detections. Preprint HUTP-85/A006 (1985)
6. R. Barbieri, M. Cerdonio, G. Fiorentini, S. Vitale, Phys. Lett. B **226**, 357 (1989)

7. F. Caspers, Y. Semertzidis, Ferri-magnetic resonance, magnetostatic waves and open resonators for axion detection, in *Proceedings of the Workshop on Cosmic Axions*, ed. by C. Jones, A. Melissinos (World Scientific Publishing Co., Singapore, 1990), p. 173
8. A.I. Kakhizde, I.V. Kolokolov, Sov. Phys. JETP **72**, 598 (1991)
9. P.V. Vorob'ev, A.I. Kakhizde, I.V. Kolokolov, Phys. At. Nucl. **58**, 959 (1995)

Chapter 18
Progress on the ARIADNE Axion Experiment

A. A. Geraci for the ARIADNE Collaboration, H. Fosbinder-Elkins,
C. Lohmeyer, J. Dargert, M. Cunningham, M. Harkness, E. Levenson-Falk,
S. Mumford, A. Kapitulnik, A. Arvanitaki, I. Lee, E. Smith, E. Wiesman,
J. Shortino, J. C. Long, W. M. Snow, C.-Y. Liu, Y. Shin, Y. Semertzidis,
and Y.-H. Lee

Abstract The Axion Resonant InterAction Detection Experiment (ARIADNE) is a collaborative effort to search for the QCD axion using techniques based on nuclear magnetic resonance (Arvanitaki and Geraci, Phys Rev Lett 113:161801, 2014). In the experiment, axions or axion-like particles would mediate short-range spin-dependent interactions between a laser-polarized ^{3}He gas and a rotating (unpolarized) tungsten source mass, acting as a tiny, fictitious magnetic field. The experiment has the potential to probe deep within the theoretically interesting regime for the QCD axion in the mass range of 0.1–10 meV, independently of cosmological assumptions. The experiment relies on a stable rotary mechanism

A. A. Geraci (✉) · C. Lohmeyer
Department of Physics and Astronomy, Northwestern University, Evanston, IL, USA

H. Fosbinder-Elkins · J. Dargert · M. Cunningham · M. Harkness
Department of Physics, University of Nevada, Reno, NV, USA
e-mail: ageraci@unr.edu

E. Levenson-Falk · S. Mumford
Department of Physics, Stanford University, Stanford, CA, USA

A. Kapitulnik
Department of Physics and Applied Physics, Stanford University, Stanford, CA, USA

A. Arvanitaki
Perimeter Institute, Waterloo, ON, Canada

I. Lee · E. Wiesman · J. Shortino · J. C. Long · W. M. Snow · C.-Y. Liu
Department of Physics, Indiana University, Bloomington, IN, USA

E. Smith
Los Alamos National Laboratory, Los Alamos, NM, USA

Y. Shin · Y. Semertzidis
IBS Center for Axion and Precision Physics Research, KAIST, Daejeon, South Korea

Y.-H. Lee
KRISS, Daejeon, Republic of Korea

© Springer International Publishing AG, part of Springer Nature 2018
G. Carosi et al. (eds.), *Microwave Cavities and Detectors for Axion Research*,
Springer Proceedings in Physics 211, https://doi.org/10.1007/978-3-319-92726-8_18

and superconducting magnetic shielding, required to screen the ^{3}He sample from ordinary magnetic noise. Progress on testing the stability of the rotary mechanism is reported, and the design for the superconducting shielding is discussed.

Keywords Axion · ARIADNE · Nuclear magnetic resonance · 3He · SQUID · Interaction · Fermions · Fifth-force · Magnetometry · Superconductor · Rotation

18.1 Introduction

The axion is a particle postulated to exist since the 1970s to explain the lack of Charge-Parity violation in the strong interactions, i.e. the apparent smallness of the angle θ_{QCD} [1, 2]. The axion is also a promising dark matter candidate. In addition, axions or axion-like particles occur quite generically in theories of physics beyond the standard model, and certain compactifications in string theory could give rise to a plenitude of axions with logarithmically distributed masses that give signatures in a wide range of experiments [3]. Thus the axion belongs to a class of "economical" and therefore highly-motivated solutions to some of the greatest puzzles in cosmology and high-energy physics. The mass of the QCD axion m_A is constrained to lie in a certain range, as shown in Fig. 18.1. The upper bound comes from astrophysics: white dwarf cooling times and Supernova 1987 A data imply $m_A < 6\,\text{meV}$ [4]. The lower bound on the axion mass depends on cosmology [12, 13]. In theories of high-energy scale inflation, the axion mass must not be lighter than about $1\,\mu\text{eV}$ to avoid its overproduction and therefore the presence of too much gravitating matter to account for the presently expanding universe. However, if the energy scale of inflation is low, depending on the initial conditions it is possible that the axion mass could be much lighter than $1\,\mu\text{eV}$, with some lower bounds resulting from black hole superradiance [14, 15]. ARIADNE [5] will probe QCD axion masses in the higher end of the traditionally allowed axion window of $1\,\mu\text{eV}$ to $6\,\text{meV}$, which are not currently accessible by any existing experiment including dark matter "haloscopes" such as ADMX [16]. Thus the experiment fills an important gap in the search for the QCD axion in this unconstrained region of parameter space.

18.1.1 Basic Principle of Experiment

The axion can mediate an interaction between fermions (e.g. nucleons) with a potential given by

$$U_{sp}(r) = \frac{\hbar^2 g_s^N g_p^N}{8\pi m_f} \left(\frac{1}{r\lambda_a} + \frac{1}{r^2} \right) e^{-\frac{r}{\lambda_a}} \left(\hat{\sigma} \cdot \hat{r} \right), \qquad (18.1)$$

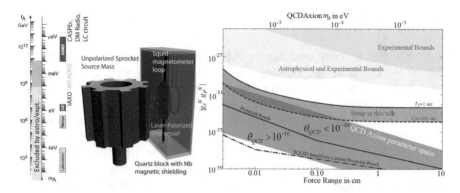

Fig. 18.1 (Left) Constraints and experiments searching for the QCD axion, adapted from [4]. (Middle) Setup: a sprocket-shaped source mass is rotated so its "teeth" pass near an NMR sample at its resonant frequency. (Right) Estimated reach for monopole-dipole axion mediated interactions. The band bounded by the red (dark) solid line and dashed line denotes the limit set by transverse magnetization noise, depending on achieved T_2 for an integration time of 10^6 s. The solid "projected reach" curve represents the sensitivity of a future, upgraded apparatus [5]. Current constraints and expectations for the QCD axion also are shown [6–11]

where m_f is their mass, $\hat{\sigma}$ is the Pauli spin matrix, **r** is the vector between them, and $\lambda_a = h/m_A c$ is the axion Compton wavelength [2, 5]. For the QCD axion the scalar and dipole coupling constants g_s^N and g_p^N are directly correlated to the axion mass. Since it couples to $\hat{\sigma}$ which is proportional to the magnetic moment of the nucleus, the axion coupling can be treated as a fictitious magnetic field B_{eff}. This fictitious field is used to resonantly drive spin precession in a sample of laser polarized cold ^{3}He gas. This is accomplished by spinning an unpolarized tungsten mass sprocket near the ^{3}He vessel. As the teeth of the sprocket pass by the sample at the nuclear Larmor precession frequency, the magnetization in the longitudinally polarized He gas begins to precess about the axis of an applied field. This precessing transverse magnetization is detected with a superconducting quantum interference device (SQUID). The ^{3}He sample acts as an amplifier to transduce the small, time-varying, fictitious magnetic field into a larger real magnetic field detectable by the SQUID. Superconducting shielding screens the sample from ordinary magnetic noise which would otherwise exceed the axion signal, while not attenuating B_{eff} [5]. The ultimate sensitivity limit is set by spin projection noise in the sample itself which scales with the inverse square root of its volume, density, and T_2, the transverse spin decoherence time [5].

The experiment can sense all axion masses in its sensitivity band simultaneously, while haloscope experiments must scan over the allowed axion oscillation frequencies (masses) by tuning a cavity [16–19] or magnetic field [20]. In contrast to other lab-generated spin-dependent fifth-force experiments using magnetometry [7–11], the resonant enhancement technique affords orders of magnitude improvement in sensitivity, sufficient to detect the QCD axion (Fig. 18.1).

18.2 Experimental Design

Detectors For detecting the effective magnetic field produced by the axion B_{eff}, three fused quartz vessels containing laser-polarized ^{3}He will serve as resonant magnetic field sensors. Three such sensors will be used to cancel common-mode noise by correlating their signals, according to the phase of the rotation of the sprocket source mass. Each sample chamber has an independent bias field control to maintain resonance between the spinning mass and the ^{3}He, during any gradual demagnetization due to the finite T_1 time. A SQUID pickup coil will sense the magnetization in each of the samples. The inside of the quartz containers will be polished to a spheroidal shape with principal axes $10\,\text{mm} \times 3\,\text{mm} \times 150\,\mu\text{m}$. The spheroidal shape will allow the magnetization to remain relatively constant throughout the sample volume, since the magnetization direction will remain reasonably well-aligned with the principal axes [21]. The cavity is fabricated by fusing together two pieces of quartz containing hemi-spheroidal cavities [22]. The vessel wall thickness is limited to $75\,\mu\text{m}$ on one side to allow close proximity to the source mass [22].

Source Mass The rotating source mass consists of a "sprocket" of height 1 cm, inner diameter 3.4 cm, and outer diameter 3.8 cm, divided into 22 sections of length 5.4 mm. The section radii are modulated by approximately $200\,\mu\text{m}$ in order to generate a time-varying potential at frequency $\omega = 11\,\omega_{\text{rot}}$, due to the difference in the axion potential as each section passes by the sensor. The factor of 11 difference between ω_{rot} and ω decouples mechanical vibration from the signal of interest. The sprocket will be driven by a ceramic shaft and precision ceramic bearings. The wobble of the sprocket will be measured in-situ using fiber coupled laser interferometers and counterweights will be applied as necessary to maintain the wobble below 0.003 cm at the outer radius. An image of a prototype tungsten sprocket appears in Fig. 18.2. Magnetic impurities in the source mass are estimated to be below the 0.4 ppm level based on measurements of similar machined material using a commercial SQUID magnetometer system [23].

Rotation Stage The cylinder will be rotated by an in-vacuum piezoelectric transducer [24] or direct-drive stage [25]. Direct drive stages [25] offer faster rotation rates (up to 25 Hz) with the caveat that local magnetic fields are larger. These can be attenuated with additional μ-metal shielding (which lies well outside of the superconducting coating on the quartz sample vessels) The rotational mechanism will need to be maintained at a higher temperature ($-30\,^{\circ}$C) than the surrounding components. This will be achieved with heaters and heat-shielding, as schematically indicated in Fig. 18.2. This design results in an expected heat load of approximately 1 W and by suitable thermal isolation we expect the excess evaporation rate of helium due to this additional heat load to be $<1.5\,\text{L/h}$.

The rotational speed must be kept constant, so that resonance between the Larmor frequency and rotational frequency can be maintained. The spin speed can be measured using an optical encoder. An index mark will allow the determination

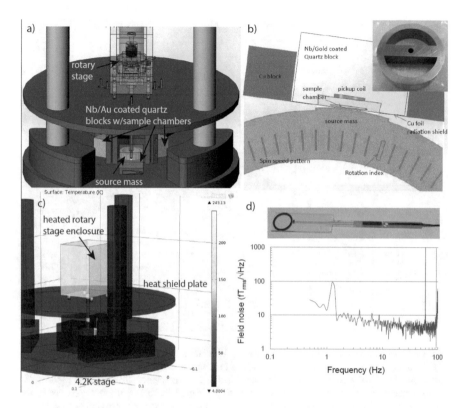

Fig. 18.2 (a) Schematic of cryostat bottom plate. The three sample regions are enclosed in quartz blocks which are coated with Niobium followed by Ti/Cu/Au to lower their emissivity. The blocks are attached to a Cu cold stage (4.2 K). A heated Au-coated enclosure houses the ultrasonic piezoelectric high-speed rotation stage, and heat shields isolate this region from the quartz block. A source mass sprocket with 11 sections is rotated around at a frequency ω_{rot}, which results in a resonance between the frequency $\omega = 11\,\omega_{\mathrm{rot}}$ at which the segments pass near the sample and the NMR frequency $2\mu_N \cdot \mathbf{B}_{\mathrm{ext}}/\hbar$, set by the magnetic field. (b) Cross sectional view from top of region near mass and detector. (Inset) prototype W rotor. (c) Thermal model of cryostat using COMSOL. Results indicate an expected He boil-off of <1.5 L/h for maintaining the motor at $-30\,^{\circ}$C. (d) (Top) prototype SQUID magnetometer fabricated on quartz substrate. (Bottom) Magnetic field sensitivity of SQUID magnetometer on Si substrate fabricated at KRISS

of the phase of the cylinder rotation, for correlation with the spin precession in the sample. Preliminary tests of the rotation speed stability of the vacuum-prepared unloaded direct-drive stage in-air indicate constant rotation speed at the part in 10^4 level at the frequency of interest with rms variation at $\sim$1 part in 3000, as shown in Fig. 18.3. This allows the sample to remain on resonance and utilize $T_2 > 100$ s.

Cryostat The samples will be housed in a liquid Helium cryostat (Fig. 18.2a). The separation between the rotating mass and quartz cell will be designed to be 75 μm at cryogenic temperature. A thermal model of the heat load appears in Fig. 18.2c. A stretched copper radiation shield foil of dimensions 25 μm × 1 cm × 1 cm will be

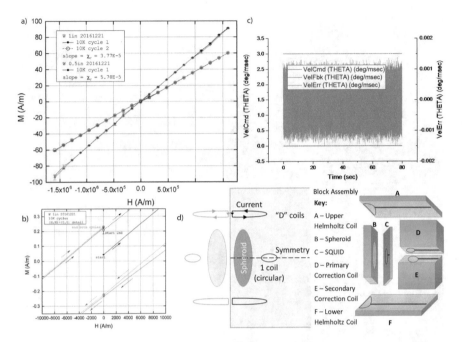

Fig. 18.3 (**a**) Magnetization test for impurities in W material used to fabricate rotor using MPMS SQUID measurement system [23]. (**b**) Zoom-in of near zero field region. (**c**) Spin-speed stability of Aerotech ADRS-100 direct drive stage unloaded and in-air as measured via an optical encoder. (**d**) (Left) A symmetrically placed coil can be used to mitigate gradients across the cell produced by the Meissner "image" of the spheroid. Two "D"-shaped coils can be used to approximate a Helmholtz coil near a superconducting boundary. (Right) Schematic of quartz block construction prior to Nb coating, showing ^{3}He sample block as well as patterned coils on additional quartz sections

inserted between the rotating mass and the quartz block. The foil will be supported by larger copper blocks affixed to the cold plate of the cryostat.

For $\omega_{rot}/2\pi = 10$ Hz, the net B_{ext} needed at the sample is of order 30 mG. B_{ext} is the sum of the internal magnetic field of the sample, which is roughly 0.2 G for 2×10^{21} cm^{-3} density of ^{3}He, and a field generated by superconducting coils. In such a field, the SQUID can operate near its optimal sensitivity of 1.5 fT/$\sqrt{\text{Hz}}$. We expect the current in the coils needs to be maintained constant at low frequencies to within $\sim$10 ppm$(1000 \text{ s}/ T_2)$.

Magnetic Shielding The fictitious field from the QCD axion coupling is at or below the 10^{-19} T level, necessitating the shielding of magnetic backgrounds. The quartz sample container will be affixed to a larger quartz block, which will be sputter coated with a 1.5 μm layer of Niobium and 200 nm of Ti/Cu. The Cu will then be electroplated with 1 μm of Gold as a blackbody reflector. The use of superconducting shielding (as opposed to e.g. μ-metal shielding) is essential to mitigate magnetic field noise from thermal currents (i.e. Johnson noise) in ordinary conducting materials. The shield also attenuates magnetic noise due to thermal currents in the tungsten mass [26], which we estimate at 10^{-12} T/$\sqrt{\text{Hz}}$, and screens the Barnett effect [27] along with the effects of magnetic impurities and the nonzero magnetic susceptibility of the source mass rotor. A magnetic shielding factor $f = 10^8$ at the nuclear Larmor frequency ~ 100 Hz would allow full design sensitivity to be attained for a T_2 of 1000 s. A list of the these requirements is shown in Table 18.1. In addition, a triple-layer μ-metal shield will enclose the cryostat while the superconducting shields are cooled through the superconducting phase transition. The total DC magnetic field should be kept below 10^{-7} T during this process to avoid "freezing-in" flux.

^3He Delivery The hyperpolarized ^{3}He gas is prepared by metastability exchange optical pumping (MEOP). MEOP is especially well-suited to the needs of this experiment: it can polarize ^{3}He at total pressures of a few mbar in an arbitrary mixture of ^{3}He and ^{4}He. The MEOP polarized ^{3}He compression system at Indiana can deliver polarized ^{3}He gas at pressures from 1 mbar to 1 bar at room temperature [28] and can therefore be used to conduct higher temperature tests in the same pressure regime that the 4 K cryogenic cell will operate in. A manifold is under

Table 18.1 Table of estimated systematic error and noise sources, as discussed in the text

Systematic effect/noise source	Background level	Notes
Magnetic gradients	3×10^{-6} T/m	Limits T_2 to ~ 100 s
		Possible to improve w/SC coils
Vibration of mass	10^{-22} T	For 10 μm mass wobble at ω_{rot}
External vibrations	5×10^{-20} T/$\sqrt{\text{Hz}}$	For 1 μm sample vibration (100 Hz)
Patch effect	$10^{-21}(\frac{V_{\text{patch}}}{0.1\text{V}})^2$ T	Can reduce with V applied to Cu foil
Flux noise in squid loop	2×10^{-20} T/$\sqrt{\text{Hz}}$	Assuming 1 μΦ_0/$\sqrt{\text{Hz}}$
Trapped flux noise in shield	$7 \times 10^{-20} \frac{\text{T}}{\sqrt{\text{Hz}}}$	Assuming 10 cm^{-2} flux density
Johnson noise	$10^{-20}(\frac{10^8}{f})$T/$\sqrt{\text{Hz}}$	f is SC shield factor (100 Hz)
Barnett effect	$10^{-22}(\frac{10^8}{f})$ T	Can be used for calibration above 10 K
Magnetic impurities in mass	$10^{-25} - 10^{-17}(\frac{\eta}{1\,\text{ppm}})(\frac{10^8}{f})$ T	η is impurity fraction
Mass magnetic susceptibility	$10^{-22}(\frac{10^8}{f})$ T	Assuming background field is 10^{-10} T
		Background field can be larger if $f > 10^8$

The projected sensitivity of the device is $3 \times 10^{-19}(\frac{1000\,\text{s}}{T_2})^{1/2}$ T/$\sqrt{\text{Hz}}$

development to deliver the gas from the polarization region into the sample area, and then to recirculate it for subsequent experimental cycles. The low spin relaxation valve technology needed to do this both at room temperature and at low temperature has been developed for neutron spin filters at neutron scattering facilities (T. Tong, private communication) and for the neutron EDM project [29].

18.3   Expected Sources of Systematic Errors and Noise

Magnetic Gradients and Frequency Shifts In order for the full sample to remain on resonance, gradients across the sample need to be controlled at the challenging level of $\sim 10^{-11} \left(\frac{1000\,\text{s}}{T_2} \right)$ T. A spheroidal chamber of uniformly magnetized gas results in a constant magnetic field in the interior. Thus the spheroidal shape of the sample suppresses magnetic gradients due to the magnetized gas itself. However, gradients and frequency shifts can be produced due to image currents arising from the Meissner effect in the superconducting shield.

A superconducting coil setup can also partially cancel the gradient, allowing extension of T_2 up to 100 s for a 99% compensation [30]. It is useful to conceptually visualize the Meissner effect as producing an spheroid of "image dipoles" on the other side of the superconducting boundary. SC coils are included in the design of the quartz block, to allow either direct cancellation (where the coil field opposes the image spheroid field at the sample) or "filling in" of the field from the image dipole (where the coil field adds to the field from the image dipole to remove gradients). We calculate that gradients can be suppressed in the central 80% region of the sample by approximately $\sim 3\times$ using the field cancellation method, or by $99\times$ using the gradient cancellation method, where the field from the image dipole is "filled in" with the coil field [30]. Two coils are included in the quartz block design so both approaches can be implemented in the experiment (see Fig. 18.3).

Bias Field Control In order to set the nuclear Larmor precession frequency at the ^{3}He sample, it is necessary to apply a constant bias field to the spheroid. The ordinary approach involving a Helmholtz coil will not work due to the close proximity of the superconducting boundary. However we can exploit the Meissner effect to our advantage by using "D"-shaped coils to create a Helmholtz like field, as shown in Fig. 18.3. Here the field from the Meissner image of the coil adds to that from the original coil to produce a nearly constant field at the location of the spheroid [30].

Acoustic Vibrations Acoustic vibrations can cause magnetic field variations due to the image magnetization arising from the Meissner effect in the superconducting shields. Although $\omega_{\text{rot}} \ll \omega$, vibrations can in principle be transmitted at ω from nonlinearities. For a $10\,\mu$m wobble in the cylinder at $\omega_{\text{rot}}/2\pi = 10\,\text{Hz}$, and 1% of this at 100 Hz, we estimate a $\delta_x \sim 2\,\text{nm}$ vibrational amplitude of the sample chamber. Assuming $\delta_x = 2\,\text{nm}$, we find that the relative motion between the sample chamber and the shield coating on the outside surface of the quartz block from elastic deformations is 10^{-17} m. With a gradient of 10^{-5} T/m, this corresponds to a field background of $\sim 10^{-22}$ T. While this background could principle be coherent, as it is associated with the rotation of the cylinder, at this level it will not dominate over any detectable axion signal for 10^6 s of averaging time.

Background vibration also should remain below $10\,\mu$m amplitude at 100 Hz, since this would produce magnetic field noise of 5×10^{-19} T/$\sqrt{\text{Hz}}$ at the resonant frequency, which can in principle begin to limit the sensitivity.

Trapped Flux   The thermal noise from a trapped flux at distance r from the sample will produce a field noise of $7 \times 10^{-20} \frac{T}{\sqrt{\text{Hz}}} \left(\frac{200\,\mu m}{r} \right)^3$ [31, 32]. This estimate is indicative and the actual noise from trapped flux will depend on the construction of the shield. In principle the experiment can tolerate a "frozen-in" DC field as large as 10^{-7} T with little deleterious effects, however the exact magnetic noise from moving flux at 100 Hz will need to be experimentally characterized for the particular shield. If necessary, additional μ-metal or cryoperm [33] shielding layers may be included inside and outside of the cryostat.

Patch Potentials   The sputtered metal films which coat the block will generally be polycrystalline, and thus will exhibit local regions of varying work function and hence local contact potential differences [34]. Fluctuating electric patch potentials can drive an oscillating time varying force on the copper heat shield membrane between the quartz sample container and the rotating mass. Assuming a 100 mV periodic signal, the force on the copper membrane will be $\sim 1.4 \times 10^{-8}$ N, resulting in ~ 5 pm of vibration. If this copper foil has an additional 100 mV potential difference with respect to the gold/Nb coated quartz, this can drive the quartz sample block with a force of 6×10^{-15} N. This in turn can vibrate the sample container's thin quartz wall with a very small amplitude of 0.5 fm. With a gradient of 10^{-5} T/m, this can cause a magnetic field background on resonance of 5×10^{-21} T. A voltage can be applied to the copper foil to minimize the DC component of such coupling, although smaller local variations can remain [35]. By slightly increasing the thickness of either the copper foil or the quartz container wall, we can increase their stiffness and thus further suppress the effect if necessary.

18.4   Discussion and Outlook

ARIADNE is a new approach towards discovering the QCD axion or axion-like particles in a mass range which is larger than that currently being probed in Dark Matter axion experiments. It is similar in style to the light shining through walls experiments such as ALPS or ALPS-II [36] in that (virtual) axions are produced in the lab, except it probes the axion scalar and dipole coupling to nuclei rather than to photons. Simulations and experimental tests conducted thus far indicate that several of the key requirements of the experiment are within reach, including the rotary-stage speed stability, thermal management for the cryostat, and magnetic gradient compensation strategy. Further experimental tests are underway with regard to the testing thin-film superconducting magnetic shielding, the ^{3}He polarization and delivery system, metrology of the masses and sample enclosures, SQUID magnetometer system, and rotational mechanism. If successful, the experiment has the potential solve the strong-CP problem and identify a particle which may make up part (or all) of the Dark Matter in the universe.

Acknowledgements We thank S. Koyu and H. Mason for computer modeling simulations at the early stages of this work. We thank T. Tong and D. Beck for helpful discussions. We acknowledge support from the U.S. National Science Foundation, grant numbers NSF-PHY 1509805, NSF-PHY 1510484, NSF-PHY 1509176. I. Lee, C-Y. Liu, J. C. Long, J. Shortino, W. M. Snow, and E. Weisman acknowledge support from the Indiana University Center for Spacetime Symmetries.

References

1. R.D. Peccei, H.R. Quinn, Phys. Rev. Lett. **38**, 1440 (1977); S. Weinberg, Phys. Rev. Lett. **40**, 223 (1978); F. Wilczek, Phys. Rev. Lett. **40**, 279 (1978)
2. J.E. Moody, F. Wilczek, Phys. Rev. D **30**, 130 (1984); P. Svrcek, E. Witten, JHEP **0606**, 051 (2006) [hep-th/0605206]
3. A. Arvanitaki, S. Dimopoulos, S. Dubovsky, N. Kaloper, J. March-Russell, Phys. Rev. D **81**,123530 (2010)
4. J. Beringer et al. [Particle Data Group Collaboration], Review of Particle Physics (RPP), Phys. Rev. D **86**, 010001 (2012)
5. A. Arvanitaki, A. Geraci, Phys. Rev. Lett. **113**, 161801 (2014)
6. G. Raffelt, Phys. Rev. D **86**, 015001 (2012)
7. G. Vasilakis, J.M. Brown, T.W. Kornack, M.V. Romalis, Phys. Rev. Lett. **103**, 261801 (2009)
8. K. Tullney, F. Allmendinger, M. Burghoff, W. Heil, S. Karpuk, W. Kilian, S. Knappe-Grneberg, W. Mller, U. Schmidt, A. Schnabel, F. Seifert, Yu. Sobolev, L. Trahms, Phys. Rev. Lett. **111**, 100801 (2013)
9. P.-H. Chu, A. Dennis, C.B. Fu, H. Gao, R. Khatiwada, G. Laskaris, K. Li, E. Smith, W.M. Snow, H. Yan, W. Zheng, Phys. Rev. D **87**, 011105(R) (2013)
10. M. Bulatowicz, R. Griffith, M. Larsen, J. Mirijanian, C.B. Fu, E. Smith, W.M. Snow, H. Yan, T.G. Walker, Phys. Rev. Lett. **111**, 102001 (2013)
11. A.N. Youdin, D. Krause Jr., K. Jagannathan, L.R. Hunter, S.K. Lamoreaux, Phys. Rev. Lett. **77**, 2170 (1996)
12. L. Visinelli, P. Gondolo, Phys. Rev. Lett. **113**, 011802 (2014)
13. D.J.E. Marsh, D. Grin, R. Hlozek, P.G. Ferreira, Phys. Rev. Lett. **113**, 011801 (2014)
14. A. Arvanitaki, S. Dubovsky, Phys. Rev. D **83**, 044026 (2011)
15. A. Arvanitaki, M. Baryakhtar, X. Huang, Phys. Rev. D **91**, 084011 (2015)
16. S.J. Asztalos, G. Carosi, C. Hagmann, D. Kinion, K. van Bibber, M. Hotz, L.J. Rosenberg, G. Rybka, J. Hoskins, J. Hwang, P. Sikivie, D.B. Tanner, R. Bradley, J. Clarke, Phys. Rev. Lett. **104**, 041301 (2010)
17. K. van Bibber, G. Carosi, *8th Patras Workshop on Axions, WIMPs, and WISPs*, Chicago, IL, July 18–22 (2012). arxiv:1304.7803
18. G. Rybka, A. Wagner, K. Patel, R. Percival, K. Ramos, A. Brill, Phys. Rev. D **91**, 011701(R) (2015)
19. B. Cabrera, S. Thomas, *Workshop Axions 2010*, University of Florida, Gainesville (2010)

20. D. Budker, P.W. Graham, M. Ledbetter, S. Rajendran, A. Sushkov, Phys. Rev. X **4**, 021030 (2014)
21. M. Tejedor, H. Rubio, L. Elbaile, R. Iglesias, IEEE Trans. Magn. **31**, 830 (1995)
22. R. Bihler, Precision Glass Blowing, www.precisionglassblowing.com
23. Magnetic Property Measurement System, Quantum Design, Inc., http://www.qdusa.com
24. Fukoku-Shinsei Corporation, www.shinseicorp.com
25. Aerotech Corporation, www.aerotech.com
26. T. Varpula, T. Poutanen, J. Appl. Phys. **55**, 4015 (1984)
27. S.J. Barnett, Phys. Rev. **6**, 239 (1915)
28. D.S. Hussey, D.R. Rich, A.S. Belov, X. Tong, H. Yang, C. Bailey, C.D. Keith, J. Hartfield, G.D.R. Hall, T.C. Black, W.M. Snow, T.R. Gentile, W.C. Chen, G.L. Jones, E. Wildman, Rev. Sci. Inst. **76**, 053503 (2005)
29. D. Beck, private communication; U. Schmidt, A. Schnabel, Yu. Sobolev, K. Tullney , Phys. Rev. Lett. **112**, 110801 (2014)
30. H. Fosbinder-Elkins, J. Dargert, M. Harkness, A.A. Geraci, E. Levenson-Falk, S. Mumford, A. Kapitulnik, Y. Shin, Y. Semertzidis, Y.-H.Lee (2017). arXiv:1710.08102
31. G.S. Park, C.E. Cunningham, B. Cabrera, M.E. Huber, Phys. Rev. Lett. **68**, 1920 (1992)
32. E. Simanek, Phys. Lett. A **194**(4), 323–330 (1994); L. Janson, Am. J. Phys. **80**(2), 133–140 (2012)
33. Amuneal Manufacturing Corp. www.amuneal.com
34. C.C. Speake, C. Trenkel, Phys. Rev. Lett. **90**, 160403 (2003)
35. R.O. Behunin, D.A.R. Dalvit, R.S. Decca, C.C. Speake (2013). arxiv:1304.4074
36. R. Bahre et al., Any light particle search ii technical design report (2013). arxiv:1302.5647

Printed in the United States
By Bookmasters